GEOLOGICAL EXCURSIONS IN THE BRISTOL DISTRICT

Jeremy

Good hunting

R. J. G. Savage

GEOLOGICAL EXCURSIONS IN THE BRISTOL DISTRICT

Edited by R. J. G. Savage

UNIVERSITY OF BRISTOL

Published by the University of Bristol

All enquiries should be addressed in the first instance to the Department of Geology, University of Bristol, Queen's Building, Bristol BS8 1TR

ISBN 0 901239 22 4

Designed and produced by the University of Bristol Information Office.
Text set in 8pt IBM Baskerville.
Headings in 10pt IBM Univers Bold.
Printed by the University of Bristol Printing Unit.

Contents

Preface

"Owing to the remarkably varied character of its rocks, the Bristol district seems specially marked out for a geological handbook."

With these words S. H. Reynolds opened the preface to his excursion guide of 1912. A glance at a geology map of Britain not only confirms this; it also shows Bristol to be the geological centre of the country as clearly as a railway map shows London to be the centre of radiating trackways. It is no accident that the father of English geology, William Smith, enunciated his principles of stratigraphy while living within a day's riding distance of Bristol. A century and a half later Bristol is still rightly a popular centre with student, amateur and professional geologists alike.

Geology has been actively studied in the University of Bristol since its foundation a century ago. Sidney Reynolds, the third professor of Geology in Bristol, wrote the first excursion handbook in 1912 with a revised edition in 1921. This book, so useful to generations of students, has long been outdated and even longer out of print. Many of the exposures described by Reynolds have since become overgrown or built upon. So much needed change that rather than attempt a further edition after so long, a completely new guide was considered necessary.

The approximate areas covered by each excursion are shown in Fig. 1 and the arrangement of the excursion is essentially stratigraphic. The geological map (Fig. 2) gives some idea of the variety of strata exposed within the region. The following is an outline of the geological systems seen on each excursion:

	Excursion
Pleistocene	16
Cretaceous	15
Jurassic	9, 10, 11, 12, 13, 14, 15
New Red Sandstone	1, 3, 9, 10, 11
Carboniferous	2, 3, 4, 5, 6, 7, 8, 9
Old Red Sandstone	2, 3, 4
Silurian	1, 4
Cambrian	1
Precambrian	1

No overall account is given of the geology of the district; this will be readily found in the *British Regional Geology Handbooks* published by HMSO. The *Bristol and Gloucester District* guide (Kellaway & Welch 1948) is the most relevant; the *Welsh Borderland* guide (Earp & Hains 1971) is relevant to **Excursion 1**, the *South-West England* guide (Edmonds, McKeown & Williams 1975) is relevant to **Excursion 2**, and the *Hampshire Basin and Adjoining Areas* guide (Chatwin 1960) is relevant to **Excursion 15.** The one-inch geological map of the Bristol District will be found very useful for many of the excursions. No fossils are illustrated as most of those mentioned in the text will be found in the British Museum (Nat. Hist.) handbooks on *British Palaeozoic Fossils* and *British Mesozoic Fossils* (obtainable through

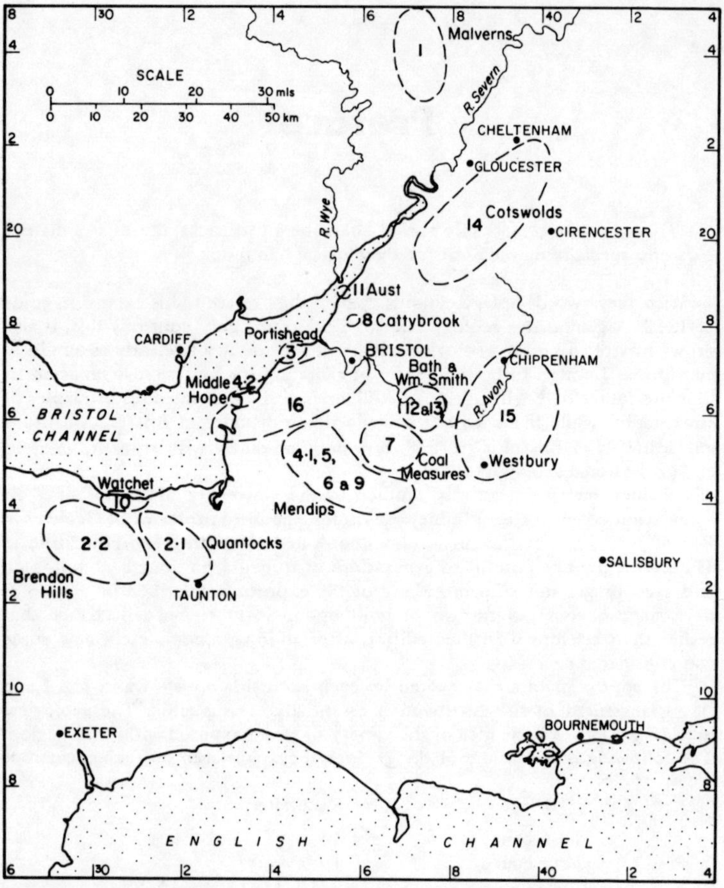

Fig. 1 Map of excursions

HMSO). The details of the itineraries reflect to some extent the author's particular interests and each author is responsible for his text. In some cases Folk nomenclature is used for carbonates; though basically a laboratory nomenclature it is becoming increasingly used for field descriptions and a key is given in **Excursion 14,** Table 2. A number of the localities are listed by the Nature Conservancy Council as Sites of Special Scientific Interest (SSSI). Those in Somerset and Gloucestershire are recorded in MacFadyen (1970). Others are in the Malverns **(Excursion 1),** and localities 2 and 3 on **Excursion 15** are in Wiltshire.

There are 183 localities detailed in the 16 excursions. Many of the 76 figures are original. The localities are mostly in Somerset and Gloucestershire, with a few in Herefordshire, Worcestershire, and Wiltshire. The new county of Avon, carved out of north Somerset and south Gloucestershire with Bristol as the county town, is not recognised on the grounds that it is still unfamiliar and its boundaries confusing for descriptive purposes. Localities in south Wales are not included as these are very ably dealt with in the recent *Geological Excursion Guide to South Wales and the Forest of Dean* (Bassett & Bassett

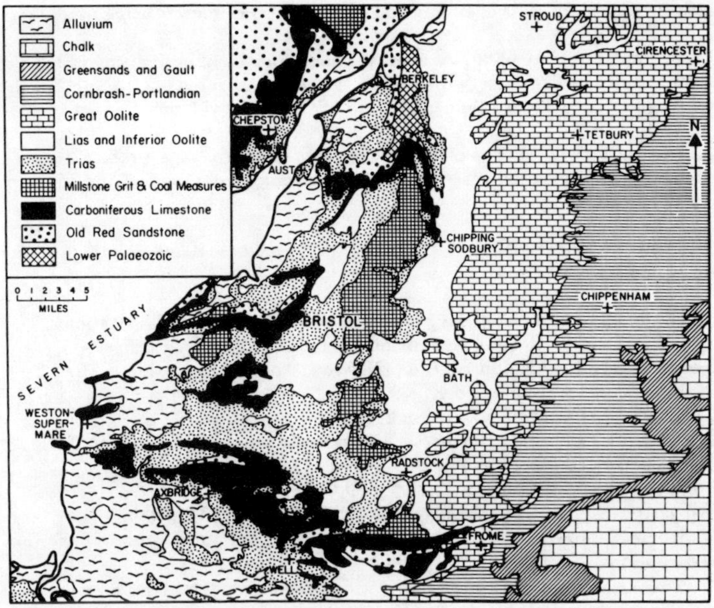

Fig. 2 A geological map of the Bristol district

1971), which can be regarded as a companion volume to this. The excursions are of varying lengths; although much depends on the time spent at each exposure, most excursions can be undertaken in a day, but **Excursion 14** would require two days. No temporary exposures are recorded, but one other omission requires explanation. The sections in Carboniferous Limestone of the Avon Gorge in Bristol have long been classic sites for geological parties. In recent years however access has become more difficult. Those on the left bank have become so overgrown that visits are virtually pointless. On the right bank access to the old quarries has been impeded in various ways and the heavy traffic on the dual carriageway (Portway) makes it very unsuitable for pedestrians.

A common pattern is used for most excursions as follows:

Objectives

The major geological phenomena to be examined on the excursion are noted.

Special Features

Information that is required before undertaking an excursion is recorded; for example where it is essential to have a low tide, where lanes make it difficult or impossible to approach sites with coaches, and in particular where permission must be obtained from owners to visit sites. Leaders of parties are advised to make a reconnaissance trip to ensure that conditions are suitable; situations can change rapidly and some of the information given here may have become outdated. Visitors are especially requested to observe the ***'Code for Geological Field Work'*** (see p. xiii).

Maps

All the maps required for complete coverage of the excursion are listed, both ordnance and geological. Both One-Inch and 1:50 000 maps are listed in this transitionary period. All localities are given six figure National Grid references, and in some cases eight figure references where the precision is required. All measurements are metric; kilometres, metres and centimetres being used.

References Listed under this heading are only the most pertinent references which the visitor may wish to have with him in the field. All references quoted in the text are included in the bibliography at the end of the volume.

Outline Geology A brief outline of the geology of the area is given as a preface to each excursion. In almost all cases a stratigraphic table and a map of the itinerary are given.

Itinerary This should **not** be undertaken until the previous items relating to the excursion have been read.

Acknowledgments Thanks are due to the following for permission to reproduce illustrations:

Bristol Naturalists' Society: Excursion 10, fig. 2; Excursion 14, figs 2–5.

David & Charles Holdings Ltd., Newton Abbot: Excursion 6, figs. 2–4; Excursion 13, figs.3b and 4.

Geologists' Association: Excursion 2, figs. 1–6.

Institute of Geological Sciences: Excursion 5, figs 3–5; Excursion 9, fig. 2; Excursion 10, figs. 1 and 3.

Linnean Society of London: Excursion 9, figs. 3 and 5.

University of Bristol, Spelaeological Society: Excursion 16, fig. 8.

Where appropriate detailed acknowledgments are to be found in the text and captions to figures.

Thanks are also recorded to Mrs Alma Gregory and Mrs Jean Bees for their draughtsmanship and to Mr. Robin Godwin for the cover photograph and to the staff of the Information Office and Printing Unit of the University for their sympathetic and friendly co-operation.

This volume has been written almost entirely by geologists who are or were on the staff of the University of Bristol (see list of contributors p. xv). Unfortunately there was no expert on the Carboniferous Limestone willing to undertake this important excursion and the Editor has reluctantly and perhaps inadequately filled the gap. We hope any omissions or errors in this volume will be brought to our attention so that they may be remedied in any subsequent edition.

Department of Geology
University of Bristol

R. J. G. Savage
April 1977

A geological code of conduct

Geological Field Work

A geological 'Code of Conduct' has become essential if opportunities for field work in the future are to be preserved. The rapid increase in field studies in recent years has tended to concentrate attention upon a limited number of localities, so that sheer collecting pressure is destroying the scientific value of irreplaceable sites. At the same time the volume of field work is causing concern to many site owners. Geologists must be seen to use the countryside with responsibility; to achieve this, the following general points should be observed.

1 Obey the Country Code, and observe local byelaws. Remember to shut gates and leave no litter.

2 Always seek prior permission before entering private land.

3 Don't interfere with machinery.

4 Don't litter fields or roads with rock fragments which might cause injury to livestock, or be a hazard to pedestrians or vehicles.

5 Avoid undue disturbance to wildlife. Plants and animals may inadvertently be displaced or destroyed by careless actions.

6 On coastal sections, be sure you know the local tide conditions. When working on coastal exposures, especially those involving cliffs, Geologists should at all times consult with the local Coastguard Service, to be made aware of possible hazards. Such advice must be respected if it bears upon the safety of a visit to a locality.

7 When working in mountainous or remote areas, follow the advice given in the pamphlet *Mountain Safety,* issued by the Central Council for Physical Education, and, in particular, inform someone of your intended route.

8 When exploring underground, be sure you have the proper equipment, and the necessary experience. Never go alone. Report to someone your departure, location, estimated time underground, and your actual return.

9 Don't take risks on insecure cliffs or rock faces. Take care not to dislodge rock, since other people may be below.

10 Be considerate. By your actions in collecting, do not render an exposure untidy or dangerous for those who follow you.

Collecting and Field Parties

1 Students should be encouraged to observe and record but not to hammer indiscriminately.

2 Keep collecting to a minimum. Avoid removing in situ fossils, rocks or minerals unless they are genuinely needed for serious study.

3 For teaching, the use of replicas is commended. The collecting of actual specimens should be restricted to those localities where there is a plentiful supply, or to scree, fallen blocks and waste tips.

4 Never collect from walls or buildings. Take care not to undermine fences, walls, bridges or other structures.

5 The leader of a field party is asked to ensure that the spirit of this Code is fulfilled, and to remind his party of the need for care and consideration at all times. He should remember that his supervisory role is of prime importance. He must be supported by adequate assistance in the field. This is particularly important on coastal sections, or over difficult terrain, where there might be a

tendency for parties to become dispersed.

Visiting Quarries

1 An individual, or the leader of a party, should have obtained prior permission to visit.
2 The leader of a party should have made himself familiar with the current state of the quarry. He should have consulted with the Manager as to where visitors may go, and what local hazards should be avoided.
3 On each visit, both arrival and departure must be reported.
4 In the quarry, the wearing of safety hats and stout boots is recommended.
5 Keep clear of vehicles and machinery.
6 Be sure that blast warning procedures are understood.
7 Beware of rock falls. Quarry faces may be highly dangerous and liable to collapse without warning.
8 Beware of sludge lagoons.
9 On visits to quarries or mines, the Health and Safety at Work Act requires that everyone on site shall wear a safety helmet. For this reason, and for more general work in geology, it is good sense to own a helmet as necessary equipment, and to wear it.

Research Workers

1 No research worker has the special right to 'dig out' any site.
2 Excavations should be back-filled where necessary to avoid hazard to men and animals and to protect vulnerable outcrops from casual collecting.
3 Don't disfigure rock surfaces with numbers or symbols in brightly coloured paint.
4 Ensure that your research material and note-books eventually become available for others by depositing them with an appropriate institution.
5 Take care that the publication of details does not lead to the destruction of vulnerable exposures. In these cases, do not give the precise location of such sites, unless this is essential to scientific argument. The details of such localities could be deposited in a national data centre for Geology.

Societies Schools and Universities

1 Foster an interest in geological sites and their wise conservation. Remember that much may be done by collective effort to help clean up overgrown sites (with permission of the owner, and in consultation with the Nature Conservancy Council).
2 Create working groups for those amateurs who wish to do field work and collect, providing leadership to direct their studies.
3 Make contact with your local County Naturalists' Trust, Field Studies Centre, or Natural History Society, to ensure that there is coordination in attempts to conserve geological sites and retain access to them.

List of contributors

ALLEN, K. C., *Department of Botany, University of Bristol*
BRADSHAW, R., *Department of Geology, University of Bristol*
COWIE, J. W., *Department of Geology, University of Bristol*
DINELEY, D. L., *Department of Geology, University of Bristol*
HAMILTON, D., *Department of Geology, University of Bristol*
HANCOCK, P. L., *Department of Geology, University of Bristol*
HAWKINS, A. B., *Department of Geology, University of Bristol*
INSOLE, A. N., *Isle of Wight Museum, Sandown, Isle of Wight*
MURRAY, J. W., *Department of Geology, University of Exeter*
SAVAGE, R. J. G., *Department of Geology, University of Bristol*
SMITH, D. Ingle, *Centre for Resource and Environment Studies, Australian National University, Canberra*
SPEEDYMAN, D. L., *Department of Geology, University of Bristol*
WHITTAKER, A., *Institute of Geological Sciences, London*
WILLIAMS, B. P. J., *Department of Geology, University of Bristol*
WRIGHT, C. A., *British Petroleum, Aberdeen*

The excursions

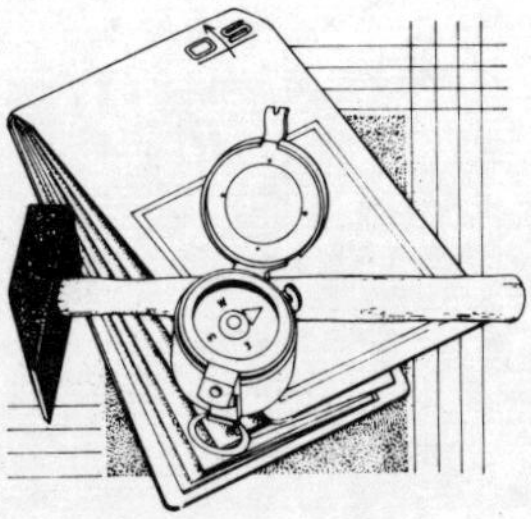

1
The southern Malvern Hills

J. W. Cowie and D. L. Speedyman

Objectives

The purpose of this excursion is to study the Precambrian of the southern Malverns, some of the younger rocks to the west, and in two localities the contact between them. The relationship between geology and topography can be appreciated at a number of points.

Special Features

Gullet Quarry and Hollybush Quarry are at present being worked, with the result that exposures of contact relationships are continually being created and destroyed. Permission to visit either quarry can be obtained from Hollybush Quarries Ltd., Ledbury, Herefordshire.

The excursion takes the form of a walk, some 10 km in length, starting at Herefordshire Beacon and following the ridge southwards to Bromsberrow just SW of Chase End Hill; coaches may be parked at both these places. If a shorter excursion (about 6 km) is required, it is suggested that localities 8 and 9 are reversed and the itinerary ended at Hollybush Quarry, where coaches may wait. Alternatively, if transport has to be left at the Herefordshire Beacon car park, the excursion can be followed as far as localities 5, 6, or 7 and a return made on foot; the transport could then be transferred to Hollybush Quarry for localities 8 and 9.

The only hostelry in the area is the British Camp Hotel.

Maps

Ordnance Survey	One-Inch Sheet 143 (Gloucester & Malvern)
	1:50 000 Sheet 150 (Worcester & The Malverns)
	1:25 000 Sheets SO 73, SO 74
Geological Survey	One-Inch New Series is not available.

Main References

Brooks (1970); Lambert & Holland (1971); Penn *et al.* (1971); Phipps & Reeve (1969).

Outline Geology

One of the few places in England where Precambrian rocks may be seen, and where there is one of the most extensive areas of English Cambrian outcrop, is in the Malvern Hills, which form a steep-sided ridge 12 km long on the border between Herefordshire and Worcestershire. The ridge itself trends N–S, and consists of Precambrian metamorphic and volcanic rocks, the break of slope closely following the limit of the Precambrian. To the west, the undulating scenery is underlain by Lower Palaeozoic rocks, while the flat plains to the east are essentially composed of Trias and Lias, with some Quaternary cover.

Research on the Malvern Hills began in the nineteenth century and a number of papers were published on the stratigraphy, structure, and petrology of the area. The history of this research and references to these papers can be found in several of the recent publications on the Malvern Hills. The *stratigraphy* and certain aspects of the geological history of the area have been described in a number of recent papers, including those by Reading & Poole (1961), Brooks & Druce (1965), Phipps & Reeve (1967), and Jones *et al.* (1969). Much controversy has arisen over the *structure* of the Malverns, particularly over the nature of the western margin of the Precambrian, and the tectonic aspects have recently been discussed by Falcon (1947), Blyth (1952),

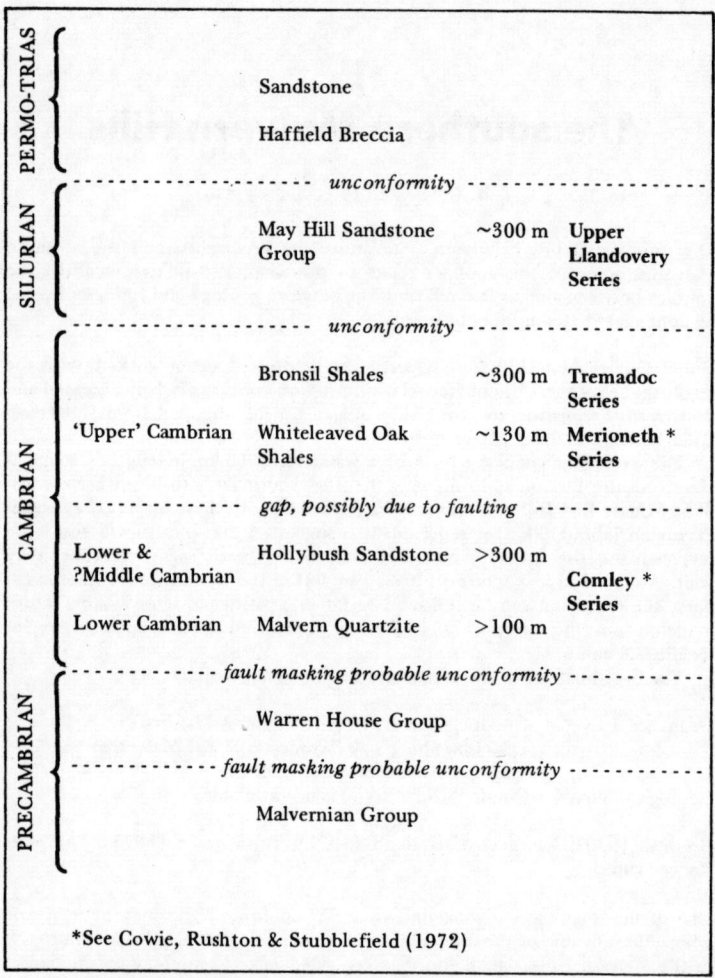

PERMO-TRIAS		Sandstone		
		Haffield Breccia		
		unconformity		
SILURIAN		May Hill Sandstone Group	~300 m	Upper Llandovery Series
		unconformity		
CAMBRIAN		Bronsil Shales	~300 m	Tremadoc Series
	'Upper' Cambrian	Whiteleaved Oak Shales	~130 m	Merioneth * Series
		gap, possibly due to faulting		
	Lower & ?Middle Cambrian	Hollybush Sandstone	>300 m	Comley * Series
	Lower Cambrian	Malvern Quartzite	>100 m	
		fault masking probable unconformity		
PRECAMBRIAN		Warren House Group		
		fault masking probable unconformity		
		Malvernian Group		

*See Cowie, Rushton & Stubblefield (1972)

Table 1 The stratigraphic succession of the rocks to be seen on the excursion

Raw (1952), Butcher (1962), Phipps & Reeve (1964, 1967, 1969), and Brooks (1970). Accounts of the *petrology and chemistry* of the Precambrian rocks have been given by Blyth & Lambert (1970), and Lambert & Holland (1971) and *geophysical surveys* have been carried out by Brooks (1968, 1969). A *Geologists' Association Guide* to the whole of the Malvern Hills has been written by Penn *et al.* (1971).

The *Malvernian Group* contains the oldest rocks in the Malvern area and consists of a metamorphic complex in which the last isotopic event took place about 600 million years ago (Lambert & Rex 1966). The rocks are very heterogeneous and range from granite to hornblendite, the most common being diorite and granodiorite. Relict igneous textures and mineralogy indicate an igneous origin for most of the rocks, although occasional garnet-mica-schists may be of metasedimentary material. The metamorphic

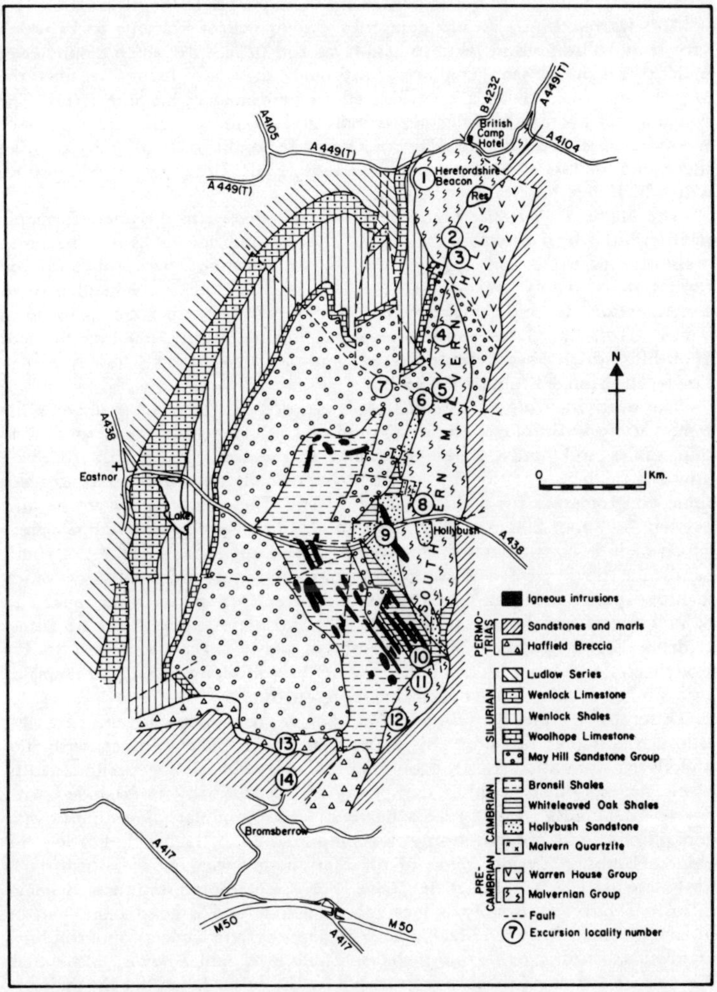

Fig. 1 Geological map of the southern Malvern Hills area

mineral paragenesis is characteristic of the greenschist or epidote-amphibolite facies. Hornblende reacts to form chlorite and epidote, biotite is chloritized, and plagioclase is sericitized and saussuritized. Shearing under moderate temperature conditions is developed in places, and this has led to the formation of a variety of schistose rocks. The intensity of the shearing generally increases from north to south, and in places has been succeeded by brittle fracturing. Some of the schists have a lamination which has previously been described as a gneissose banding, but which is produced by the segregation of chlorite and mica into layers under strong shearing conditions. There are no true high-grade metamorphic gneisses in the Malvernian complex. 'Banding' is also produced by the injection of narrow acid pegmatites and granitic veins along the schistosity. The complex is cut by a number of minor intrusions, mainly microdiorite dykes, and these post-date the shearing.

The *Warren House Group* comprises a sequence of volcanic rocks which vary from spilitic pillow lavas to acid lavas and tuffs, into which a number of basic dykes have been emplaced. Although they are faulted against the Malvernian, these rocks are considered to be younger than the latter. The Precambrian age of the volcanics is indicated by the presence of pebbles of Warren House type in the Malvern Quartzite at the base of the Cambrian succession in this area. They closely resemble the Uriconian of Shropshire, with which they have been correlated.

The *Malvern Quartzite* occurs in small fault-slices and the unconformable relationship which probably exists with the Precambrian rocks is not seen in present exposures, and the original thickness cannot be estimated. The formation is mainly grey quartzite which is sometimes cross-bedded, with conglomeratic developments showing pebbles, mainly of quartz, ranging up to 5 mm across in places. Fossils recorded include horny brachiopods and hyolithids which have definite Lower Cambrian associations (Comley Series: Cowie, Rushton & Stubblefield 1972).

The overlying *Hollybush Sandstone* is exposed over large areas and exhibits some variety in lithology. About 20 m of greenish glauconitic sandstones with thin shales and sandy, shaly limestones at the base is probably followed upwards in the succession by grey and green sandstones with quartzites and some conglomerate for a thickness of 110 m. These latter beds are in turn overlain by about 200 m of sandstone, coloured shades of green and pale grey, which include conglomerates. Only the basal and upper parts of the Hollybush Sandstone have yielded fossils; a brachiopod genus and hyolithids which includes species found in the Malvern Quartzite. This sparse fauna appears to be of Lower Cambrian age (Comley Series) and there appears to be no faunal evidence for assigning a Middle Cambrian age (St. David's Series) to the Hollybush Sandstone. By comparison with Shropshire, however, there remains a possibility of this younger age being represented.

Outcrops of the *Whiteleaved Oak Shales* are comparatively extensive although usually obscured by superficial cover. The contact with the underlying formation is probably faulted everywhere. The predominantly black shales, which weather dark grey, are occasionally interbedded with coarse glauconitic quartzitic sandstones and nodular limestones with concretions, and contain horny brachiopods and ostracods indicating the Upper Cambrian *Olenus* Zone of the Merioneth Series (Cowie, Rushton & Stubblefield 1972). The shales have been metamorphosed near igneous intrusions but this is only a local phenomenon and a good fauna can be obtained which includes brachiopods (*Broeggeria* and *Orusia*) and trilobites (agnostids, *Peltura, Sphaerophthalmus, Ctenopyge,* and *Lakella*). More than one zonal horizon is probably represented by this latter fauna but the majority of the fossil species can be assigned to the *Peltura scarabaeoides* Zone of the Upper Cambrian.

The comformably overlying *Bronsil Shales* are of Tremadoc age and vary in colour from bluish-grey to olive or yellow, and can thus be easily distinguished from the much darker underlying formation; characteristic features are partings and rusty brown weathering. The lower beds characteristically yield the dendroid graptolite *Dictyonema* which is associated in the upper beds with trilobites. The shales form fertile low-lying ground so that exposures are few and fossils comparatively hard to obtain – a remark which can be applied equally to most of the Cambrian beds of the Malverns.

Following an important period of non-deposition and erosion in the area, sedimentary rocks of the upper part of the Silurian Llandovery Series were deposited. These belong to the *May Hill Sandstone Group,* and Groom (1910) recognized three formations: Woolhope Shale Formation; Wyche Formation;

Cowleigh Park Formation. In this excursion the Cowleigh Park Formation only will be seen. These are characteristically brown to maroon or purple interbedded sandstones and siltstones near the base, succeeded above by conglomerates, sandstones, and minor shales (Phipps & Reeve 1967). The thickness is about 120 m in the southern Malverns.

The Lower Palaeozoic strata are unconformably overlain by Mesozoic strata; the basal beds are the conglomeratic *Haffield Breccia* and these are succeeded by cross-bedded red *sandstones* both being Permo-Trias.

Itinerary

Locality numbers refer to those on the maps, Figs 1–4.

1 **Herefordshire Beacon** (SO 750 400) There is a large car park on the south side of the A 449 opposite the British Camp Hotel (SO 763 403), but coaches must park on the gravel lay-by on the opposite side of the road, by the First Aid Station. Take the footpath from the car park south-westwards to the top of Herefordshire Beacon, which is made of Precambrian hornblende-diorite of the Malvernian Group. Note the small reservoir on the eastern side of the Beacon, beyond which are rounded hills of Precambrian volcanic rock belonging to the Warren House Group. On the summit of Herefordshire Beacon (S0 750 400) are the extensive earthworks of an ancient hill fort, known as British Camp, dating from the second century BC. From the summit a magnificent view of the Malvern ridge and surrounding countryside may be obtained, and a close relationship between topography, vegetation, and geology can be discerned. To the north the ridge of Malvernian rocks continues, but is offset by faulting. Silurian rocks form the undulating scenery immediately to the west and NW of the ridge; the Woolhope Limestone and Wenlock Limestone form tree-clad ridges, and broad folds, which by mapping can be shown to be overturned in places, can be determined from the arcuate forms of some of these ridges. Towards the SW the Obelisk (SO 7523 3783) stands on Silurian May Hill Sandstone, beyond which the lower ground is occupied by Cambrian Bronsil Shales. The Precambrian ridge continues southwards, and to the east lies the flat plain of Triassic rocks, which have been faulted against the Precambrian in many places, but occasionally rest unconformably on the latter.

2 **Direction Indicator** (SO 762 395) Follow the crest of the Precambrian ridge southwards for 0.5 km to the southern rise of the hill fort complex, passing over intermittent outcrops of Malvernian on the path. Beyond the rise the path continues along the crest of the ridge, and then shortly makes a sharp turn to the left, descending the eastern slope at the southern end of the fort. As the descent is made, look towards the east for a squat stone pillar 0.5 m high at the centre of a broad saddle about 100 m away, and head towards it. This pillar is a direction indicator and is partly made of blocks of various Malvernian rocks; note the narrow granite vein in the central block. It is situated on the faulted boundary between the Malvernian to the west and the Warren House volcanics to the east; the fault passes northwards through the reservoir and southwards along a broad gully.

3 **Clutter's Cave** (SO 7628 3938) Follow the prominent southward-leading path along the west flank of the ridge (the path is indicated on the pillar as leading to Giant's Cave). Just over 100 m along the path, on the left, lies Clutter's Cave (or Giant's Cave), formed in spilitic lavas of the Warren House Group. The Warren House volcanics comprise spilites in the west, with acid and intermediate lavas and tuffs to the east (Lambert & Holland 1971), and have been correlated with the Uriconian of Shropshire. At Clutter's Cave poorly-developed pillow structures can be discerned in a few places. These are best displayed over the entrance to the cave and also just to the right of the cave entrance on a level with the top of the arch. Here, pillows up to about 25 cm across sag down between the underlying pillows. In these

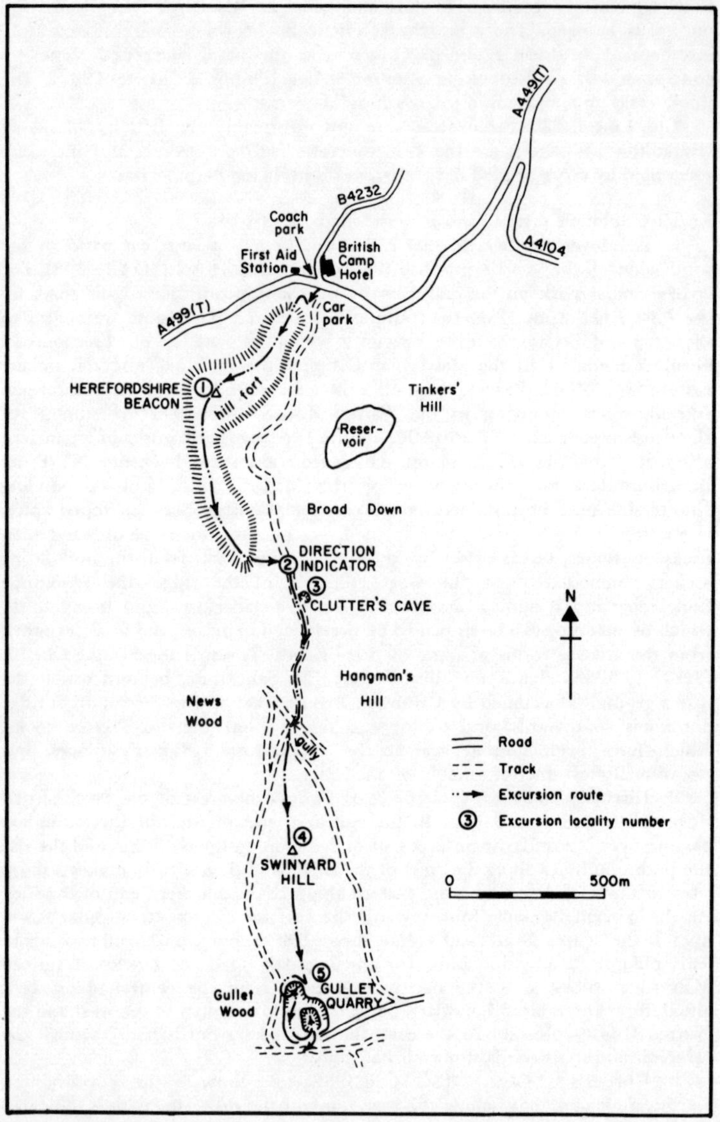

Fig. 2 Localities 1–5

spilites both the clinopyroxene and the sodic plagioclase form microscopic laths, and there are numerous small amygdales filled with chlorite.

4 Swinyard Hill (SO 762 386) From Clutter's Cave follow the path southwards along the west flank of Hangman's Hill, descending slightly all the time. After 0.4 km the path bears right and crosses a fault gully trending NW–SE, south of which Malvernian rocks outcrop again. Follow the path along the crest of the ridge to the top of Swinyard Hill. There are intermittent exposures of Malvernian along the whole length of the crest of Swinyard Hill, the lithology varying from microcline-granite to epidote-amphibolite which is often veined by the granite. Continue past the top of the hill along the crest of the ridge, where occasionally fine-grained garnet-mica-schists apparently of metasedimentary origin are seen within the Malvernian, until Gullet Quarry is seen dropping away immediately to the west of the path on the far side of a fence. Stop here and go westwards, skirting around the northern margin of the quarry. Start to descend through the woods on the west side of the quarry, bearing to the left at the earliest opportunity in order to arrive at the uppermost level in the extreme NW corner of the quarry.

5 **Gullet Quarry.** The steep contact between Precambrian, to the east, and Silurian passes through this quarry. It is exposed best in the extreme NW of the quarry at the uppermost level (SO 7618 3818). The Precambrian consists of Malvernian quartz-chlorite-muscovite-schists, the lamination of which dips at 50° towards 070°. In thin-section the platy minerals are seen to be greatly deformed. The Silurian sandstones and shales of the May Hill Sandstone Group dip at about 60° towards an average azimuth of 275° near to the contact, but at angles as low as 20° towards the same direction further from the contact. The May Hill Group is of Upper Llandovery age and is here quite richly fossiliferous. Near the subsoil at the top of the quarry the beds are overturned and dip to the east at about 35°. This phenomenon may be due to superficial gravitational bending ('hill creep') or perhaps was caused by ice-drag, but the possibility of dragging along a fault should not be ruled out. The contact relationships are complex. The boundary between the Silurian and Precambrian is irregular, and the contact zone has been hematitized, particularly along surfaces dipping at about 70° towards 260°. Fossiliferous Silurian shales penetrate down into clefts in the Precambrian, and these shales have been hematitized in parts, with coral moulds being preserved in hematite. What is seen at this exposure depends upon the progress of quarry working, but an irregular surface with a conglomerate containing large Precambrian pebbles and boulders has been seen between Precambrian and Silurian rocks. Detailed descriptions of the contact relationships have been given in a number of recent papers, but the nature of the contact has been in some dispute. It has been interpreted as an unconformity by Reading & Poole (1961), Butcher (1962), and Brooks & Druce (1965), but as a fundamental Hercynian fault by Phipps & Reeve (1964). Proponents of the fault theory maintain that the upturning of the Silurian strata towards the contact in Gullet Quarry is a result of the faulting. Brooks (1969) backs up the unconformity theory with geophysical evidence. His interpretation of the geophysical data is that the Precambrian basement lies at a shallow depth beneath the Lower Palaeozoic rocks to the west of the Malvern Hills, and he concludes that there cannot be a major western boundary fault of fundamental character.

Descend the track on the western side of the quarry, looking back towards the contact just studied and noticing its southern continuation along the western margin of the quarry.

If time permits, descend to the very lowest level of the quarry. The rocks exposed here are essentially heterogeneous granodioritic schists with chloritic and amphibolitic mafic layers, the latter probably represent metamorphosed doleritic intrusions. Lamination is well displayed in places, but the rocks are massive and structureless in others. The lamination is defined in the field both by variable lithologies and by mineral segregation within lithological units. In thin-section it is seen that fine-scale lamination is produced by the alternation of layers of chlorite and mica with layers of quartz and feldspar. Often there is evidence of deformation along the chlorite and mica layers, and occasionally garnets in these layers are shattered by this deformation. The lamination is

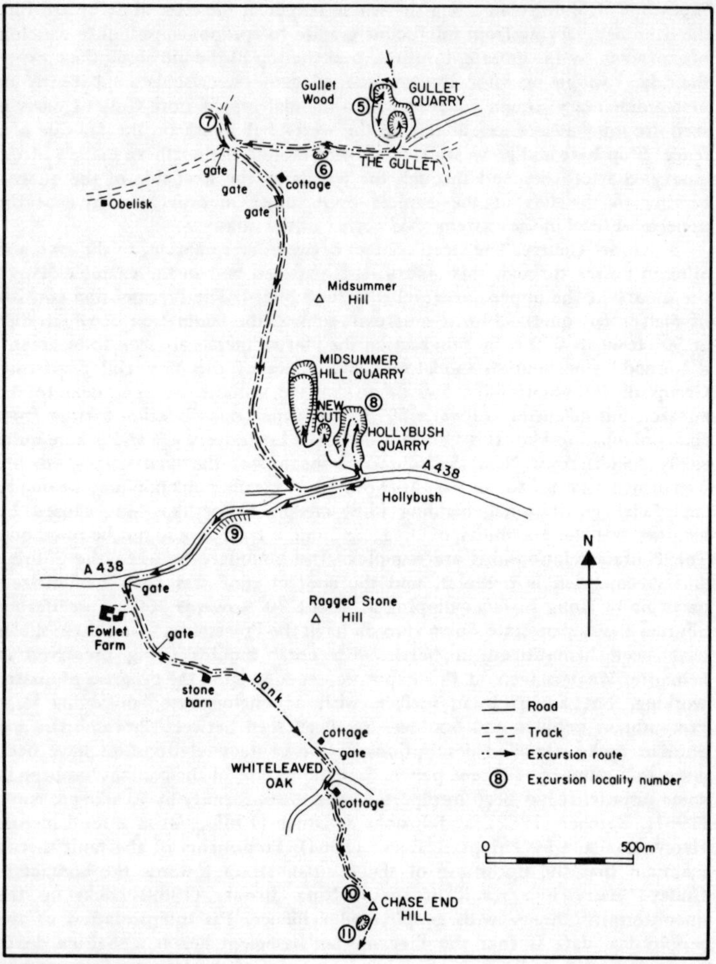

Fig. 3 Localities 5–11

steeply-dipping, and in places has been gently folded. The direction of dip of the lamination varies; on the east side of the quarry it dips at about 70° towards 125°, whereas on the west it dips at about 70° towards 170°. The strike of the lamination here is very different from that at the contact with the Silurian described above, and in places the strike is perpendicular to the N–S trend of the Malvernian outcrop. Pinkish microcline-chlorite-granite veins are very common, and these are generally parallel to the lamination. They vary in size from pegmatitic sheets a metre or so in width, down to narrow veins a few millimetres wide giving the rock a banded appearance. Faulting and shearing is much in evidence here, and may be associated with The Gullet E–W trending fault immediately to the south. There has been a certain amount of mineralization along joint planes, and the minerals which have crystallized include calcite, epidote, and pyrites.

6 'The Gullet' From Gullet Quarry entrance turn westwards and follow the lane up a gentle rise through 'The Gullet', a narrow E–W trending fault valley. After 200 m, just round a slight left bend, there is a small digging, about 5 m square, just off the lane on the left (SO 7600 3798). This is a small exposure of Cambrian Malvern Quartzite, which is a quartzitic sandstone containing conglomeratic horizons with abundant quartz pebbles 8–10 mm in size. The pebbles are mainly rounded, but some angular fragments also occur. Horny brachiopods of the obolid type can usually be found, particularly in the upper beds, and have proved a Lower Cambrian (Comley Series) age. The dip is 35° towards 320°. The relationship here between the Cambrian and Precambrian is uncertain; it may be an unconformity or a fault.

7 Laneside exposure (SO 7565 3812) Continue up the lane for about 300 m until a gate is seen ahead where the lane bends to the right in front of the gate (SO 7564 3801). Do not go through this gate, but follow the lane to the right immediately in front of it and continue for approximately 120 m. On the west side of the path at this point, about 2 m above the path and 3 m from it, is a small exposure. The rock is Cowleigh Sandstone of the May Hill Group, and is a fairly massive, well-jointed, muddy sandstone containing decalcified brachiopods and crinoid stems. It is brown on fresh surfaces, and also weathers brown. The dip is 40° towards 290°. The exposures of this May Hill Sandstone continue along the west side of the path for about 20 m.

Retrace steps to the gate mentioned above (SO 7564 3801), and again do not go through the gate but take the path which goes towards the SE, passing through a different gate. Follow this path for about 300 m, passing a cottage which lies below on the left, to a second gate. Go through this gate and follow the track, bearing slightly to the right, along the western flank of Midsummer Hill. Continue down the track for 1 km until it meets the A 438 at Hollybush. Turn left and follow the A 438 eastwards for 200 m to the entrance to Hollybush Quarry on the left.

8 Hollybush Quarry (SO 761 371). In the main part of the quarry, on the lowest level, the rocks are purple, green, and cream coloured heterogeneous Malvernian granodiorite, with chloritic and amphibolitic bands and granite veins. Most of the mafic minerals have been altered to green chlorite, sometimes with calcite. Green epidote is a common secondary mineral, forming both from green hornblende and as a result of the saussuritization of plagioclase. The banding, when discernible, dips steeply towards the NNW and there is a great deal of deformation and shearing of the rocks in places. Veins of calcite, epidote, and hematite are common, and are often associated with shears.

Take the track which leaves the western side of the quarry and ascends to the higher quarries just to the NW. After the first sharp bend to the right, a small quarry known as the **New Cut** (SO 7595 3707) is seen ahead, situated immediately above the main quarry which lies below to the east. Most of the New Cut consists of Malvernian granodiorite with a microcline-leucogranitic facies along part of the western side. As the western side of the Cut is ascended, the rocks become strongly foliated and brecciated, with the foliation dipping steeply to the NNE. Shearing is in evidence, and in thin-section crystal straining and the development of a mortar texture witness to the crushing undergone by these rocks. There is considerable mineralization by calcite and hematite, the latter now much altered to limonite. These minerals occur both as irregular net-veins and as diffuse patches permeating the rocks.

In the rubble-covered slope leading out of the western side of the Cut are occasional small outcrops of glauconitic Cambrian Hollybush Sandstone. Jones *et al.* (1969) describe the Malvern Quartzite as underlying the Hollybush Sandstone on the western side of the New Cut, and although there is evidence of shearing, they maintain that there is no major faulting. The morphology of the New Cut, however, is liable to change as quarrying progresses.

From the New Cut, follow the track westwards into the lower level of Midsummer Hill Quarry. On the western side of the narrow entrance to the quarry the faulted contact between Malvernian schists and Cambrian sediments can be seen (SO 7590 3710). Traversing on towards the quarry, the first rocks at the beginning of the entrance are folded ferruginous Hollybush sandstones and shales. These are followed by a 2 m zone of highly contorted shales and marls, including a grey mudstone. Along the fault itself the rocks are greatly sheared and fractured, beyond which is Malvernian granodioritic schist. The fault trends N–S, and continues along the eastern face of the quarry, its course being marked by the presence of weathered hematite along the fault plane. To the west of the fault the rocks are Malvernian, and to the east Lower Palaeozoic rocks continue from the New Cut. The younger rocks are therefore down-faulted to form a Lower Palaeozoic outlier, surrounded by the Malvernian.

The fault is also exposed in the far NE corner of the quarry, particularly at the top level, but constant working of the quarry is continually destroying exposures of this contact or making them inaccessible, while creating new ones. Jones *et al.,* however, have described a fossiliferous Upper Llandovery limestone adjacent to the fault here, the limestone containing blocks of Hollybush Sandstone at its base, and being separated from the sandstone by a thin seam of sheared clay. Jones *et al.* (op. cit.) maintain that although there has been shearing, the relationship between the Silurian and the Cambrian is an unconformable one.

9 Roadside exposure (SO 756 368) Return to the main Hollybush Quarry, and at the entrance turn right and walk westwards along the A 438. After 0.4 km there is a broad left-hand bend in the road where rocks are exposed on the inside of the bend. The exposure is about 100 m long, and most of the rock is trachyte which forms a dyke intruded into slightly folded and faulted early Cambrian Hollybush Sandstone. It is better to traverse this section from west to east.

At the western end, well-bedded greenish Hollybush Sandstone is gently folded and has a general dip to the WNW. Early Cambrian fossils may be found in this sandstone, which in thin-section is seen to be a micaceous protoquartzite. The characteristic glauconite of the Hollybush Sandstone which is found elsewhere is poorly developed here. Towards the east the bedding becomes less evident as the dyke is approached, although there is no evidence in thin-section of marked hornfelsing by the intrusion. The contact between the massive fine-grained sandstone and the greenish-grey fine-grained marginal facies of the dyke occurs on the crown of the bend, but is not readily discernible. However, it appears to dip steeply towards the west. The lithologies look somewhat similar, but the sandstone generally contains sporadic sparkling flakes of muscovite. There are veins of calcite in the contact region, but these are not parallel to the dyke margin.

Near the contact the trachyte is fine-grained and greenish-grey, but further to the east, towards the centre of the intrusion, it becomes medium-grained and has a pinkish or purplish hue. The trachyte is an unusual type, and is porphyritic, consisting mainly of altered potassium feldspar, often coloured pinkish-brown, and very small phenocryst pseudomorphs of calcite, green chlorophaeite, quartz, and iron ore, apparently after amphibole. Chlorophaeite also occurs in amygdales, and in the marginal faces of the intrusion is common as irregular aggregates in the groundmass, giving the rock its greenish-grey colour. The feldspar laths have a trachytic texture in which flow layers diverge around phenocrysts.

At the extreme eastern end of the exposure the igneous rock becomes finer-grained again, and the other dyke margin can, with some difficulty, be located. The easternmost rocks exposed are of greenish-grey, slightly glauconitic, micaceous Hollybush Sandstone. The width of the dyke as measured in this section is about 60 m. The fine-grained margins are about 10 m wide, and their contacts with the central part of the dyke are diffuse, and appear to dip westwards. A large number of minor igneous intrusions of Ordovician age have been emplaced into the Cambrian shales to the south and west of this locality, and this dyke is probably one of the Ordovician suite.

Continue westwards along the A 438 for 0.4 km, passing a cottage on the left, and going round a slight bend to the right. At (SO 7528 3657) turn left through a gate, along a farm track. Do not follow the main track as it turns to the right into Fowlet Farm, but keep straight on, passing to the east of the farm. Follow the minor track straight ahead and after 150 m pass through another gate and bear left. The ground to the right is occupied by Bronsil Shales (Tremadoc Series of the Cambrian), the tree-clad ridges being formed of Ordovician igneous intrusions. As the track heads towards the Precambrian ridge of Ragged Stone Hill it crosses the Silurian May Hill Sandstone. The Obelisk, seen away to the left, is also on the May Hill Sandstone Group, but this formation outcrops in disconnected outliers and there are numerous faults in between as well as unconformable relationships.

200 m after the gate mentioned above, keep on the track which passes immediately adjacent to the northern side of a stone barn alongside a ditch. (The track by the stone barn tends to become rather muddy in wet weather.) Follow along the south side of the ditch and when, after 100 m, the ditch peters out continue along its line, keeping to the SW side of the bank. The path now traverses the Whiteleaved Oak Shales (Merioneth Series of the Cambrian) which occupy the lower ground to the SW. Whale-back ridges in the field to the right of the path are occupied by igneous intrusions. Small exposures of igneous rock and black shale of the country rock, in places bleached white by the intrusions, can be found with some searching. Loose material of this nature is probably close to its original source bed-rock. Continue along the SW side of the bank to the gate by a cottage in the bottom SE corner of the field. Go through the gate and follow the track about 100 m to the road in the hamlet of Whiteleaved Oak.

10 Whiteleaved Oak. At the road turn right, and after 40 m turn left along a track which passes adjacent to the right-hand side of a cottage. The track shortly degenerates into a footpath which continues between hedgerows for 200 m. Black Whiteleaved Oak Shales floor the path in many places, and on the margins of the path under the hedges can occasionally be found, with some searching, fragments of metamorphosed shales which have been baked and bleached by the proximity of Ordovician igneous intrusions. After a left-hand bend the path emerges into open ground. Just round this bend, on the right of the path at ground level is a small, geologist-eroded exposure of dark Whiteleaved Oak Shales (SO 7602 3567). It contains calcareous nodules which can yield Cambrian (Merioneth Series) fossils, mainly brachiopods.

11 Chase End Hill (SO 761 355) Follow the grassy path, across the break of slope which marks the boundary between the Cambrian and the Precambrian, to the summit of Chase End Hill which consists of Malvernian diorite and granodiorite. From here, there is a magnificent view of the Precambrian ridge as it continues northwards on the far side of the Cambrian shales upon which the hamlet of Whiteleaved Oak is situated. Towards the NW the tree-clad intrusions in the Whiteleaved Oak Shales and the Bronsil Shales can be seen, and to the south the Precambrian ridge comes to an end in the Triassic plain. Beyond the Trias to the east stretches the broad Vale of Evesham lying on Lower Lias, traversed by the River Severn passing over Quaternary deposits, with the Cotswold Hills in the distance. To the ENE Bredon Hill stands up as an outlier of the Cotswold rocks. Follow the path

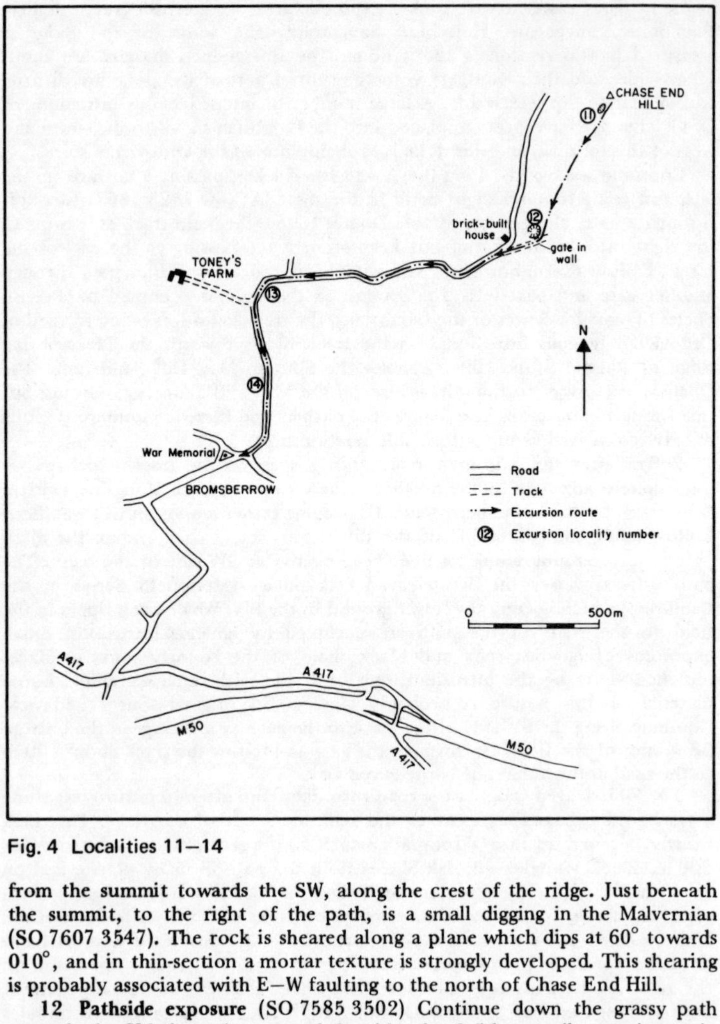

Fig. 4 Localities 11–14

from the summit towards the SW, along the crest of the ridge. Just beneath the summit, to the right of the path, is a small digging in the Malvernian (SO 7607 3547). The rock is sheared along a plane which dips at 60° towards 010°, and in thin-section a mortar texture is strongly developed. This shearing is probably associated with E–W faulting to the north of Chase End Hill.

12 Pathside exposure (SO 7585 3502) Continue down the grassy path towards the SW along the crest of the ridge for 0.5 km, until a gate is seen in the wall on the left where the path bears to the right. Below the path at this point and to its right is a large partly-overgrown digging in which there are numerous exposures of Malvernian diorite and granite (SO 7585 3502). In places the rocks are strongly sheared, the plane of shearing here dipping at 40° towards 080°, which is approximately perpendicular to that at the previous locality, and may be associated with a N–S trending fault along the west side of Chase End Hill.

13 Toney's Farm Do not go through the above-mentioned gate in the wall, but follow the path round to the right in front of the gate. Continue on this path for 100 m until it meets a lane by a brick-built house; turn left onto the lane. Continue along the lane for 1 km until the road makes a sharp left-hand bend shortly before Toney's Farm. Just round this bend there are a few outcrops of Permo-Triassic Haffield Breccia in the steep banks on either side of the lane (SO 748 347).

14 **Bromsberrow** Follow the lane past the entrance to Toney's Farm and continue southwards towards Bromsberrow. In the banks at the sides of the lane, particularly on the right, are good exposures of Permo-Triassic Sandstone, the best outcrops being just past the slight bend to the right (SO 7482 3440). The millet-seed grains and dune-bedding are worthy of note. Continue down the lane towards the hamlet of Bromsberrow, and when after 200 m the junction with a slightly wider road is reached, turn right. This leads immediately to the War Memorial in Bromsberrow which stands in the centre of a large grassy triangle where a coach can turn and park (SO 748 341). (The road to Bromsberrow from the A 417 is negotiable by coach. The direct road between Hollybush and Chase End Street is not suitable for coaches.)

2
The Quantock and Brendon Hills, West Somerset and North Devon

D. L. Dineley

Objectives This section concerns the Devonian rocks which outcrop in the south-western limits of the Bristol region and which form the attractive uplands to the south of Bridgwater Bay. The itineraries cross picturesque farmlands and moorland with many of the steep hillsides covered by Forestry Commission plantations. The region is one of historical and archaeological interest as well as of importance, geologically and it has experienced scattered but not inconsiderable mining, largely for iron ore in the sixteenth to early twentieth centuries.

Special Features Most of the localities listed are small and virtually all are on private property. Local residents and landowners are sympathetic to geologists who ask permission to cross land or visit exposures. The region has, however, suffered from inconsiderate parties who do not exercise such courtesy. Most of the lanes are unsuitable for coaches.

Permission is required to visit the following localities:
Cannington Park Quarry: Castle Hill Quarry Co., Cannington, Bridgwater
Triscombe Quarry: Anglo American Asphalt Co., Dene Court, Bishop's Lydeard, Taunton
Treborough Quarry: West Somerset District Council, Minehead

Maps

Ordnance Survey	One-Inch Sheets 164 (Minehead), 165 (Weston-super-Mare) 177 (Taunton & Lyme Regis)
	1:50 000 Sheet 181 (Minehead & Brendon Hills)
	1:25 000 Sheets SS 93, SS 94, ST 03, ST 13, ST 14 ST 22, ST 23, ST 24
Geological Survey	One-Inch Sheets 294 (Dulverton), 295 (Taunton)

Main References Webby (1965a); Webby (1965b); Webby & Thomas (1965).

Outline Geology Geological interest in the area was fostered by the early work of De la Beche, and of Sedgwick and Murchison in the first half of last century. W. A. E. Ussher and others carried out the survey for the Geological Survey's Taunton Sheet (New Series) almost a hundred years ago but only Webby has in recent years published detailed studies of the structure and stratigraphy of the region.

The oldest rocks present are Middle (or? Late Lower) Devonian grits, sandstones and slates, intricately folded, cleaved and faulted prior to the deposition of the New Red Sandstone. Later Mesozoic rocks occur just beyond the present localities and it may be presumed that they once were present unconformably overlying the Palaeozoic strata. Late Tertiary denudation is thought to have been largely responsible for the present high

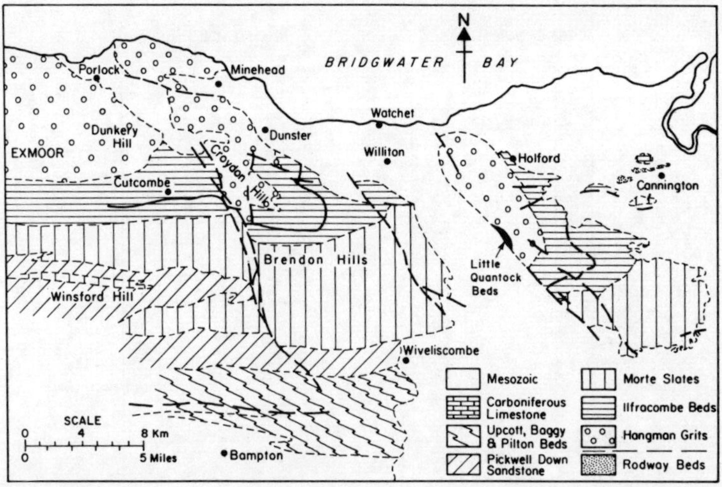

Fig. 1 Geological map of the Brendon and Quantock Hills (after Webby & Thomas 1965)

level erosion surfaces with deepening of the narrow valleys, or combes, by periglacial activity during Quaternary times.

The classical division of the Devonian rocks of West Somerset and North Devon is:

Pilton Beds	-------------
Baggy and Upcott Beds	
Pickwell Down Beds	Upper Devonian
Morte Slates	
Ilfracombe Beds	-------------
	Middle Devonian

Hangman Grits	
Lynton Beds	Lower Devonian

Representatives of all but the lowest and the uppermost two divisions are known in the Brendon and Quantock Hills. The Hangman Grits vary relatively little throughout their outcrop in North Devon and West Somerset but the Ilfracombe Beds, Morte Slates and Pickwell Down Beds are represented by local units distinct from those westward along the strike. Webby's (1965a) stratigraphic units for the Brendon Hills have been accepted by the Institute of Geological Sciences (Edmonds, McKeown & Williams 1969). Some differences exist between the successions in the Brendons and the Quantocks and these are regarded (Webby 1965b) as the result of transgressive movements of the Mid-Devonian shoreline.

Hangman Grits Only the upper part of the Hangman Grits of the North Devon coastal section occurs in the region, giving rise to the high land in the northern Quantocks, to the south of Dunster and at Dunkery Hill. The Grits are predominantly buff to maroon, massive and well jointed quartzitic sandstones. Individual beds are of uneven thickness, generally between 2 cm

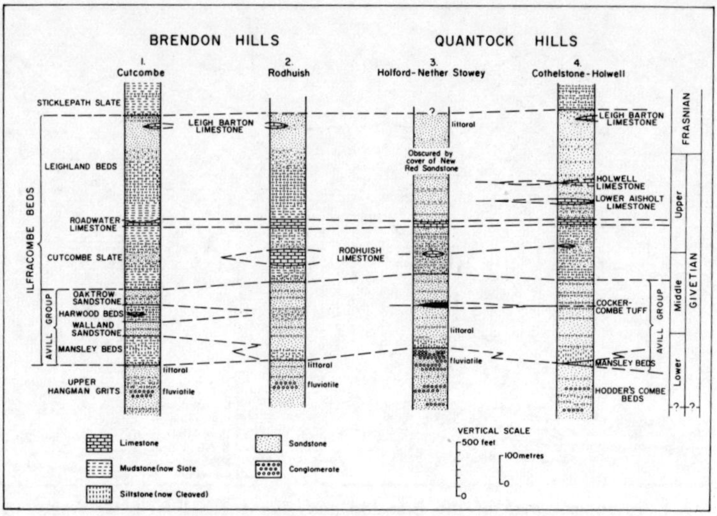

Fig. 2 Geological sections in Brendon and Quantock Hills (after Webby 1965)

and 3 m. Thin interbeds of cleaved siltstone and slate become more abundant near the junction with the overlying Ilfracombe Beds. Rare internal structures include lamination and cross-lamination with rare trace fossils. Some bedding surfaces may be rippled. Conglomeratic layers are uncommon except in the north. Rare fossils include the bivalves *Nuculites* sp., *Nuculoidea* sp. and *Spathella* sp. and linear plant fragments resembling *Psilophyton.*

In the Quantocks Webby (*op. cit.*) recognised two subdivisions of the Hangman Grits, the *Triscombe Beds* below and the *Hodders Combe Beds* above. The lower formation is present between Triscombe and Quantoxhead and includes light green sandstones with minor brown siltstone, sandstone, green mudstone, intraclast conglomerates and rare quartz-pebble conglomerates. About 500 m of this unit is estimated east of Triscombe, but the thickness may increase to the north.

The *Hodders Combe Beds* are somewhat similar but include more, thicker, quartzitic sandstones and massive conglomerates. They reach some 315 m in thickness and may be correlated with the Upper Hangman Grits of the Brendon Hills or the Rawn's Beds and the Sherry Combe Beds of North Devon. The only fossils, however, are poorly preserved plant fragments.

The *Little Quantock Beds* are a distinctive group of brown and greyish siltstones and slates found in the vicinity of Triscombe (ST 156 356), and contain *Chonetes* sp., *Tentaculites* sp., and crinoid ossicles. Ussher regarded them as down-faulted Ilfracombe Beds but Webby (1965c, p. 322) interpreted them as Hangman Grits and as perhaps the oldest exposed strata in the Quantocks.

Ilfracombe Beds The Ilfracombe Beds are clearly a rather variable division lying conformably above the Hangman Grits. Webby mapped four main units which include further subdivisions (see Webby *op. cit.*, Fig. 1). Correlation of these with the Ilfracombe strata on the coast is still uncertain. The Avill Group is the oldest unit and in the Brendons has the following subdivisions:

The *Mansley Beds* are dark grey slates interbedded with thin grey siltstones, rare rusty brown siltstones containing *Chondrites,* and a few massive sandstones.

The *Walland Sandstone* is a unit of grey-brownish, unfossiliferous, massive sandstone with minor bands of slate and siltstone, in total 10 m thick.

The *Harwood Beds* are well-bedded grey slates, with minor cleaved calcareous siltstones and sandy limestones near the mid-part of the unit. Fossils include *Spirocyrtia ascendens* (Spriestersbach) *Schuchertella;* sp., *Nuculoidea?* sp., *Ctenodonta* cf, *aequis* Spriestersbach, *Nuculites dewalquii* (Asselberghs), *Actinopteria* sp., *Myophoria* sp., *Platyceras (Platyceras) compressum* (F. A. Roemer) and *Cupularostrum?* sp.

The *Oaktrow Sandstone,* a local unit of interbedded sandstones and siltstones, which has a similar fauna to that just listed for beds below but also includes *Thomasaria gibbosa* Vandercammen, *Athyris?* sp., *Cranaena* sp., *Centronella* sp., *Aviculopecten?* sp., *Carydium* sp., *Plethomytilus* aff. *villmarensis* (Frech), other gastropods, bryozoa and crinoids.

In the Quantock Hills the subdivisions of the Avill Group (Fig. 2) are not easily recognised. Most of the outcrop is of undifferentiated cleaved siltstones and slates with a few massive sandstones. Fossils are locally common. An element not seen in the Brendons is a green lithic tuff, the Cockercombe Tuff containing marine fossils. Exposures at Holford (ST 155 412) may be of the same bed.

The *Cutcombe Slate* is throughout the area predominantly formed of grey and brownish slates with calcareous siltstone lenses. Towards the middle of this formation arenaceous beds and siltstones occur together with the thin Rodhuish Limestone, a somewhat slaty thin-bedded to massive limestone. Thick crinoidal beds are conspicuous in this limestone and there is a rich coral fauna, including *Heliophyllum halli* Milne-Edwards & Haime, *Thamnophyllum caespitosum* (Goldfuss), *Disphyllum aequiseptatum* (Milne-Edwards & Haime), *Thamnopora* aff. *cronigera* (d'Orbigny), *T. reticulata* (de Blainville), *'Cystiphyllum' secundum* (Goldfuss) and stromatoporoids.

The *Roadwater Limestone* is an important, persistent and distinctive marker band of pinkish grey crystalline limestone with a fauna of corals, brachiopods, bryozoa, crinoids and gastropods. The extensive coral list includes *Heterophrentis percevali* Webby, *Heliophyllum halli, Thamnophyllum caespitosum, Disphyllum aequiseptatum, Acanthophyllum concavum* (Walther), *Digonophyllum bilaterale* ? (Champernowne), *Mesophyllum sandhillense* Webby, *'Cystiphyllum' secundum, Thamnopora cervicornis, T. polyforata* (Schlotheim), *Alveolites suborbicularis* Lamarck, together with brachiopods, lamellibranchs, polyzoans and crinoid ossicles.

The *Leighland Beds* include both slates and sandstones with three fossiliferous limestones present locally. Fossils are elsewhere rare and poorly preserved. The two lower limestone bands (the Lower Aisholt Limestone and the Holwell Limestone) are confined to the Quantocks but the upper limestone (the Leigh Barton Limestone) occurs also in the Brendon Hills as a cleaved unit of crinoidal slate and crystalline limestone containing brachiopods and bryozoa.

Morte Slates Between 1200 m and 1700 m of somewhat uniform unfossiliferous slates and cleaved siltstones represent the Morte Slates. In the Brendon Hills the lowest division of the Morte Slates (the *Sticklepath Slate*) is formed of grey smooth and silty slate with rare fine sandstones up to 10 cms thick. In the Quantock Hills the slates are maroon and grey with a few massive sandstones; many are cross-laminated.

Rodway Beds These are flaggy, micaceous, maroon unfossiliferous siltstones and sandstones which are thought to be the youngest Devonian rocks of the area. They occur only north of the A 39 east of Nether Stowey and seem to be uncleaved and less strongly indurated than the Morte Slates to the west.

Igneous Rocks Minor intrusive igneous bodies occur at the Combe (ST 269 291) and at Hestercombe (ST 243 292). Those at the Combe are thin

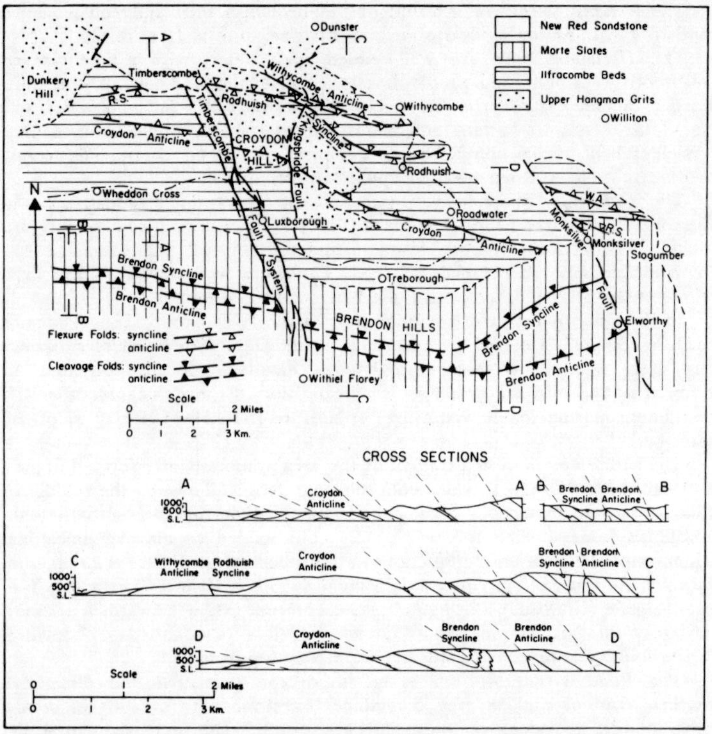

Fig. 3 Geological structures in the Brendon Hills (after Webby 1965)

sheets of quartz-diorite and postdate the folding and cleavage. At Hestercombe the mass is a diorite of uncertain shape. The origins of these rocks are uncertain.

Geological Structure

The Devonian rocks of North Devon and West Somerset in general dip to the south at moderate to high angles and are affected by cleavage and high-angle faulting. The Brendon Hills lie within the main Devonian outcrop. The Quantocks are separated from them by a graben-like structure of New Red Sandstone and are part of a broad anticlinal structure striking northwest. It has been thought possible that these Palaeozoic masses rest upon a major thrust which may reach the surface not far north of the coast westwards from Bridgwater Bay; no thrust has, however, been discovered.

Webby *(op. cit.)* distinguishes two kinds of fold structures in these rocks. In all the main stratigraphic divisions minor folds, *cleavage folds,* are found and predate the *flexure folds.* The axial planes of these folds dip in southerly directions at between 20° and 70° and tend to plunge to the east at somewhat lower angles. Many cleavage folds are overturned. The cleavage itself trends east to ENE and dips at between 30° and 70° to the SSW in the Brendons and in the southern part of the Quantocks. To the north the trend is mainly ENE and the dip commonly 40°.

The larger, *flexure,* folds affect the width and position of outcrop of the stratigraphic divisions and include the following:

The *Courtway Anticline,* a broad, open, east-plunging flexure extending

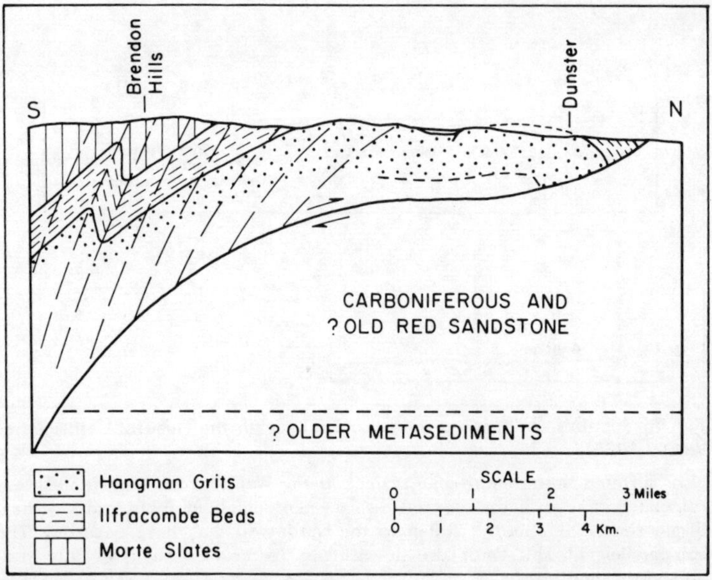

Fig. 4a Section to illustrate the structure beneath the Brendon Hills (after Webby 1965)

ESE across the centre of the Quantocks.

The *Croydon Anticline,* another broad, open symmetrical flexure, plunging gently to the east at between 10° and 15°. This structure extends west across the higher ground of Monkham Hill, Blackhill and Rodhuish Common towards Croydon Hill: west of the Timberscombe Faults it continues west to die out north of Mansley Combe.

The *Rodhuish Syncline,* an open flexure trending more or less parallel to, and situated about 2 km north of, the Croydon Anticline. It plunges at between 10° and 20° to the east. At Timberscombe the syncline passes under the New Red Sandstone to reappear locally on the eastern side of Dunkery Hill.

The *Withycombe Anticline,* a minor structure on the southern flank of Withycombe Hill. It trends east, plunging gently in that direction as far as the New Red Sandstone.

Important faults occur within both the Brendon Hills and the Quantocks Devonian and all have a strong NW-N trend. They are younger than the folds. In the Brendon Hills the *Timberscombe Faults* are perhaps the most conspicuous in topographical results and appear to have been initiated in pre-New Red Sandstone times. Further movement has occured after the deposition of those red rocks. Both horizontal and vertical displacement has taken place with a total of about 450 m vertical downthrow to the west, and a dextral displacement of about 1,100 m.

The *Kingsbridge Fault* runs south between Croydon Hill and Monkham Hill and appears to involve a vertical displacement only, downthrowing to the west amounts from about 110 m in the north to 26 m in the south.

The *Monksilver Fault* extends SSE along the valley from Yard (ST 075 366) and on towards Elworthy (ST 083 349).

In the Quantock Hills the *Gothelstone Fault* controls the western margin of the Devonian outcrop. Originating in pre-New Red Sandstone times, it has

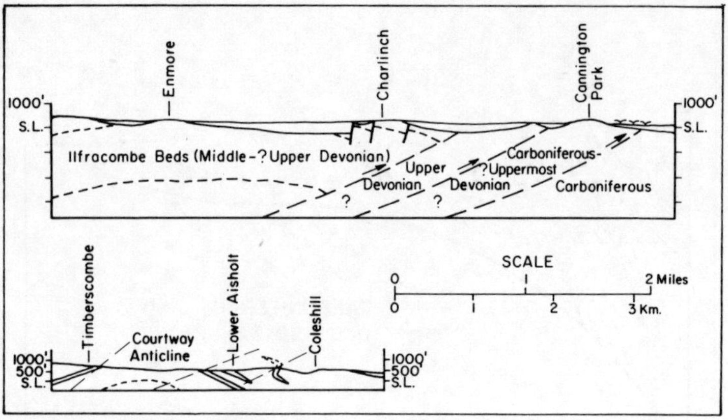

Fig. 4b Sections to illustrate the structure beneath the Quantock Hills (after Webby 1965)

also suffered minor movement since then. Webby (*op. cit.,* p.339) has calculated that an actual dextral displacement of about 6 km and a vertical displacement of about 2 200 m to the south-west may have occurred. The sub-parallel Gib Hill Fault also downthrows to the west but is of only local and minor significance.

It has long been recognised that the Cannington Park inlier of much-deformed Carboniferous (Viséan) Limestone must lie adjacent to a structural break of sufficient size to cut out much of the local Devonian succession. Several authors have postulated that a thrust underlies it and geophysical evidence – a fall in gravity values from north to south between the coast and the north slope of the Quantocks – supports the contention. Webby (1965c, p.340) showed that on geological grounds no thrust can be projected between Cannington Park and the Rodway (Devonian) inlier. He suggested that a simple fault with a displacement of perhaps less than 1 000 m would suffice to explain the outcrop geology. To accommodate the geophysical data, however, he imagined that a major thrust could occur to the north of the Cannington Park inlier and that two smaller breaks may be placed between the inliers to the south (see Fig. 4b). If, as Webby (1965c) suggested and as is mentioned above, the flexure folds of the Brendon Hills represent broad warps on the crest of a major thrust, the *Courtway Anticline* may occupy a similar position to that of the *Withycombe Anticline* in the west.

Itinerary 1: The Quantock Hills and Local Palaeozoic Inliers

Beginning at the quarry in Carboniferous Limestone at Cannington Park this excursion includes exposures on the eastern side of the Quantock Hills followed by localities on the western and northern flanks of the Hills. The route requires transport by car if all localities are to be visited in a single day.

1 **Cannington Park Quarry** (ST 251 404) Cannington Park lies between the villages of Cannington and Combwich, on the west bank of the River Parrett. There are two quarries, the southern one is the working quarry and the northern one (ST 246 407) currently houses the quarry plant. In Cannington village take the road north (Rodway) signposted Hinkley Point; after 800 m the working quarry will be seen on the right and the second quarry is 600 m further on.

Macroscopic fossils are not common nor easy to extract from the matrix. They include: *Koninckophyllum praecursor* Howell (most common); *Lithostotion martini* Milne-Edwards & Haime; *Syringopora distans* Fischer;

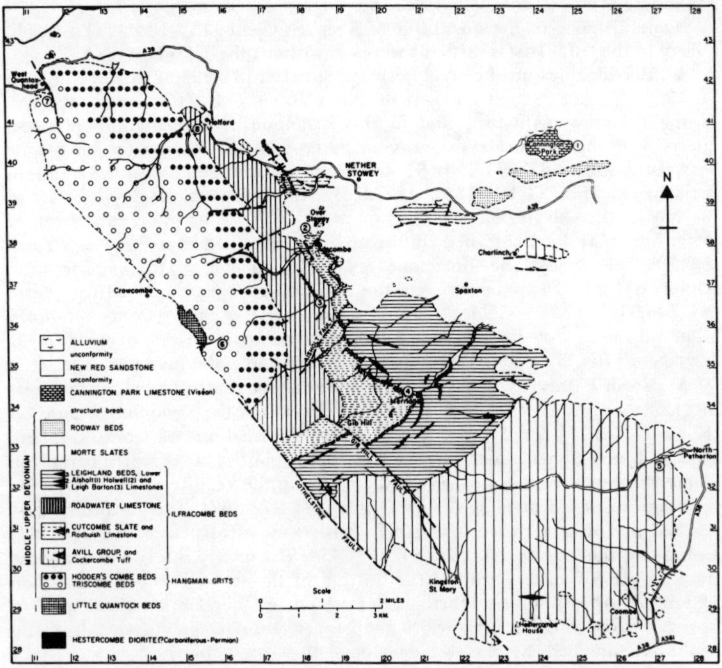

Fig. 5 Geological map of the Quantock Hills (after Webby 1965)

Productids of the hemisphericus group; *Seminula ficoidea* Vaughan. The age of these beds is thought to be Visean, the Upper C to S1 zone of the Avonian section (Wallis 1924).

2 **Over Stowey Area** Approximately 2 km south of Nether Stowey, Quarry Breach (ST 178 386) lies on the lower eastern flank of the Quantock Hills and close to the unconformity of the New Red Sandstone. The locality includes an old quarry a few metres to the north of the road from Nether Stowey to Crowcombe and 2 km SW of Nether Stowey. Slates and siltstones here are part of the Avill Group, dipping north east at between 55° and 60°.

The old quarry on the south side of the road at Adscombe (ST 186 380) is very overgrown but may show massive, recrystallised pinkish grey limestone beds belonging to the Roadwater Limestone and containing: *Syringaxon* sp., *Temnophyllum inflatum* Walther; *Alveolites* cf. *suborbicularis* Lamarck; *Thamnopora reticulata* (de Blainville); *Coenites medius* Lecompte; stromatoporoids and crinoid ossicles. The Roadwater Limestone thus includes fossils indicative of an upper Middle Devonian (Givetian) age.

3 **Cockercombe** (ST 187 366) Cockercombe lies in Quantock Forest, about 3.5 km south of Nether Stowey, and stretches westwards across the Devonian succession from the Cutcombe Slate in the lower part of the combe to the Hodder's Combe Beds in the upper reaches. Exposures are small and vary from year to year with the state of the lane and stream banks. A good exposure can be seen behind the Forestry Commission cabin. The laneside exposures are for the most part confined to the Avill Group slates and sandstones, dipping NW at between 25° and 50°. This lane is the type locality of the Cockercombe Tuff which occurs about 230 m above the base of the Group. It is a green lithic tuff with fragments of spilite and with fine bands

containing bryozoa, crinoid ossicles and indeterminate brachiopods.

Some distance to the north side of Keeper's Combe (ST 183 371) is an old quarry in the tuff. This is difficult to reach without the 1:25 000 map.

4 **Merridge** lies on the road between Spaxton (ST 225 370) and Kingston St. Mary (ST 222 296) at a height of about 200 m O.D. It is close to the axis of the Courtway Anticline and in this neighbourhood many old pits and quarry workings delineate the curving outcrop of the Roadwater Limestone between Aisholt (ST 194 356) SE to Merridge and then SW to the southern flank of Gib Hill (ST 195 335). At Gib Hill the limestone outcrop is offset to the NW by the Gib Hill Fault. There are many local, rather poor, exposures, of which two may be mentioned: in the quarry in the field opposite Torr's Farm (ST 208 344) crystalline limestone yields *Hexagonaria* aff. *quadrigemma* (Goldfuss) and Thamnopora sp.; the old quarry west of Headford Farm (ST 206 345) has yielded *Syringaxon* sp.; *Endophyllum abditum* Milne-Edwards *Alveolites* sp.; *Thamnopora reticulata* (de Blainville); *T. ?cervicomis* (de Blainville); *Coenites medius* Lecomple and bryozoa.

5 **North Petherton** From Merridge the road runs eastwards along the crest of Quantock Forest to North Petherton, ascending the Devonian section on the way. Near Lydeard Cross (ST 231 330) the road crosses from Leighland Beds to Morte Slates which occupy the slopes about as far as North Petherton where the New Red Sandstone cuts them out. In the vicinity of this village the best exposures are at King's Cliff Quarry (ST 277 367). This locality is best reached by the minor road from North Petherton to Thurloxton (ST 275 304) and taking the lane to Boomer (ST 277 329). The quarry is a 100 m along the track to the SW of the bend at the bottom of the slope. It shows more than 100 m of unfossiliferous Morte Slates which are here interbedded grey to brown cleaved siltstones, brownish sandstones and dark grey shales. In many of the siltstone beds there is well-developed cross-lamination.

6 **Triscombe Quarry** (ST 161 356) From Hestercombe the itinerary now turns northwards to visit exposures in the central and western parts of the Quantock Hills. The road may be followed through Fulford (ST 210 294) to join the A 358. Triscombe is approximately 1 km to the east of the main road and the large quarry lies 5 km further east up the minor road. Here the Triscombe Beds, the lower of the two undoubted coarse clastic subdivisions of the Hangman Grits, outcrop as light green, massive quartzitic sandstones with minor green mudstone intraclast conglomerates. Many of the beds are laminated or cross-laminated and some of the sandstones contain rare plant remains. The rocks dip at about 20° ESE.

Triscombe hamlet lies upon the outcrop of Little Quantock Beds but no exposures worthy of study are available. In this area the contact of the New Red Sandstone with the underlying Devonian is an unconformity according to Webby (1965c) but the Geological Survey map (Sheet 295) shows the Cothelstone Fault locally forming part of this junction. No clear evidence of faulting has been found in recent years.

7 **West Quantoxhead** From Triscombe the lane via Crowcombe may be taken to the main (A 358) road and West Quantoxhead reached via Bicknoller. The steep hill slopes are on Hangman Grits and the change in slope marks the position of the unconformity of the New Red Sandstone. There are three quarries at West Quantoxhead, all in Triscombe Beds (ST 113 415). The entrance is by the lane on the east side of the Windmill Inn car park. All of these exposures are dangerous in that the sandstones are closely jointed and falls of blocks from the quarry faces are frequent.

The western quarry may be reached from the lane ascending Beacon Hill. It shows massive and well-bedded brown quartzitic sandstones with minor siltstone and slate together with rare bands of quartz pebble and mudstone intraclast conglomerate. The beds dip north at about 20°. A few linear plant fragments are poorly preserved in the sandstones. The middle and eastern quarries show much the same beds but the conglomeratic bands are not present. The lithologies and sedimentary structures of these beds are similar to those of the Old Red Sandstone and may be fluviatile in origin.

8 Holford The A 39 eastwards from West Quantoxhead skirts the high ground of the Hangman Grits outcrop while to the north the New Red Sandstone extends to the coast. Holford lies on the NE corner of the Devonian outcrop and upon the line of the unconformity itself. The lane section south of the church (ST 156 409) is now rather overgrown but exposures of the Avill Group here are of reddish slates and silty brown quartzitic sandstones, dipping NE at about 25°. In the past these beds have yielded casts and moulds of thick-shelled bivalves and gastropods: *Myalina* ?sp.; *Edmondia* ?sp.; *Spathella* sp.; *Pedasiola* cf. *rhenana* Spriestersbach; *Bellerophon* ?sp.; *Euryzone* ?sp. and *Spinocyrtia* ?sp. The fossils tend to be disarticulated and are thought to represent an intertidal littoral accumulation.

In Holford Glen near the junction with Hodder's Combe (ST 155 412) are further exposures of the Avill Group in the stream bed. They are of bedded quartzitic sandstones and lithic sandstones on top of which, and dipping at about 20° to the NE, is a band of maroon weathered tuff. At the top of the tuff is a thin band containing *Cyrtospirifer* sp., *Atrypa* ?sp., *Nuculoidea* sp., abundant bryozoa and crinoid ossicles. In the stream bed the sandstones above the tuff are faulted against gently dipping New Red Sandstone conglomerates and shales. An unconformable contact between the two formations can be seen, however, on the higher western banks of the stream. In Hodder's Combe (ST 151 408), at a disused quarry, there are beds high in the Hangman Grits. In detail they are like those at the West Quantoxhead quarries.

Itinerary 2: The Brendon Hills

This excursion provides a chance to see the formations at the eastern end of the Devonian outcrop in North Devon. Many of the roadside exposures are somewhat overgrown but the maintenance of the roads and their banks (and cuttings) provides many temporary exposures. Although the Hangman Grits form the highest ground, they are not widely exposed inland. Good sections in the Hangman Grits can be seen, however, in the cliffs between Minehead and Hurlstone Point (ST 899 493).

The highest point in the region, Dunkery Beacon, may be reached from the A 39 road via Luccombe (SS 911 445) or via the A 396 from Dunster. The route followed here begins at Oaktrow in the Avill Valley, south-west of Dunster.

1 Oaktrow Leaving Dunster on A 396, the road south from Pitt Bridge (SS 943 413) almost to Wheddon Cross (SS 924 388) ascends the steep-sided Avill valley in which many local and commonly indifferent exposures occur.

Harwood Beds are exposed in the old quarries on the eastern side of the road (SS 939 405). They comprise calcareous siltstones and impure limestone bands. The beds are folded into an easterly plunging anticline and the well-marked cleavage dips SSE at about 30°. Fossils found in these calcareous beds include: *Spinocyrtia ascendens*? (Spriestersbach); *Cupularostrum* ?sp.; *Schuchertella*? sp.; *Nuculoidea* sp.; *Actinopteria* cf. *rudis* Phillips; bryozoa, shell fragments and crinoid ossicles.

In a small quarry in Oaktrow sandstone on east side of A 396, 100 m north of turning to Oaktrow Farm (SS 939 401) sandstone beds approaching the massive are folded into a minor syncline which is overturned so that the southern limb dips south at about 50°. The plunge is to the south. The thickening of the individual beds at the apex of the fold is conspicuous.

The small quarry 50 m to the south of previous stop has a deep hollow in one wall and is referred to by Webby & Thomas (1965, p.187) as the 'cave quarry'. It exposes massive calcareous sandstones overlain by thin reddish

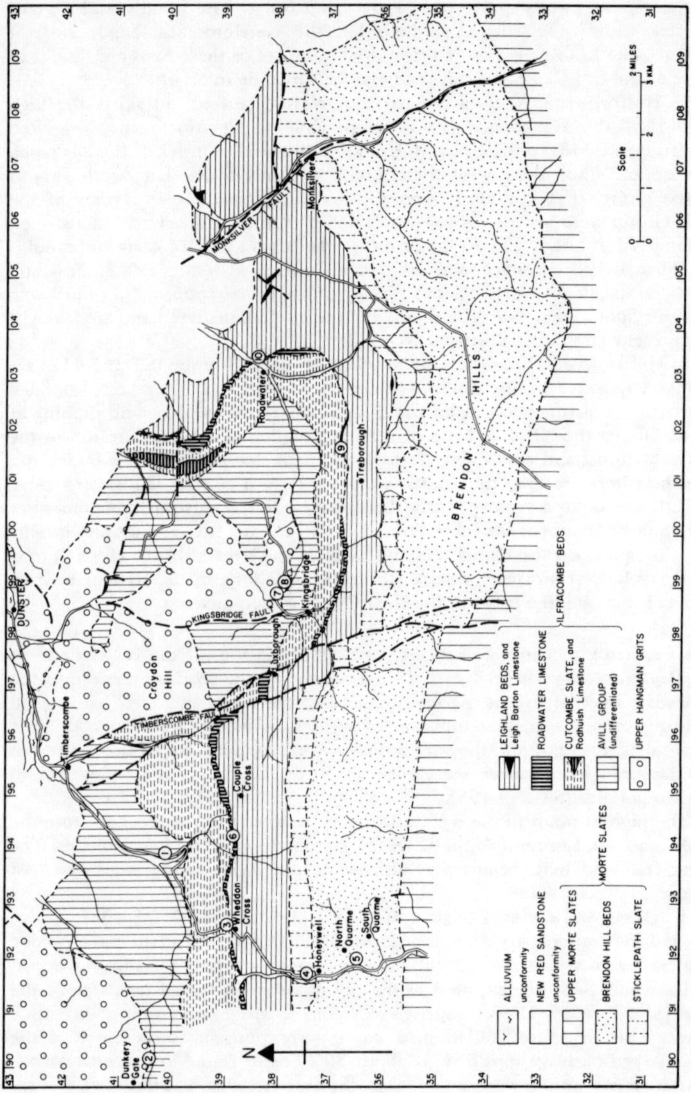

Fig. 6 Geological map of the Brendon Hills (after Webby 1965)

brown fossiliferous rotten stone a few centimetres thick and cleaved siltstones. The beds above the sandstone have yielded: *Spinocyrtia ascendens*, *'Spirifer'* sp., *Thomasaria gibbosa* Vandercammen, *Cytrina* sp., *Cupularostrum*? sp., *Athyris*? sp., *Schuchertella* sp., *Cranaena* sp., *Centronella* sp., cf. *rudis*, *Aviculopecten*? sp., *Carydium* sp., *Myophoria*? sp., *Plethomytilus* aff. *villmarensis* (Frech), *Platyceras compressum* (F. A. Roemer), other gastropods, polyzoa and crinoid ossicles.

2 Dunkery Gate and Mansley Combe (ST 896 406) From the A 396 at Wheddon Cross the B 3224 may be taken to Dunkery Gate. Dunkery Beacon is within easy walking distance and provides an excellent viewpoint of the area. Hangman Grit exposures, however, are few and poor near the Beacon. To the west of the road at Dunkery Gate the stream exposes a few typical maroon to brown quartzitic sandstones and the top of the Hangman Grits. On the eastern side of the road at Dunkery Gate, Mansley Brook begins its steep descent through Mansley Combe. It exposes a good section in grey and brownish finely laminated shales and cleaved siltstones dipping southwards at about 50°. The bedding planes and open, weathered joints are extensively covered with yellowish limonite here, the origin of which is uncertain. The siltstones are up to 5 cm thick and may contain *Chondrites* sp. Throughout much of this section the bedding is less steeply dipping than the cleavage.

3 **Wheddon Cross** (ST 925 388) In the cuttings behind Normans Cottages on the NE side of the main (A 396) road exposure in the uppermost Cutcombe Slate and lowest Leighland Beds may be seen. Permission to examine these exposures should be sought from the tenants. The rocks exposed are grey and brown slates with rare siltstones dipping south at about 40°. There are numerous thin white quartz veins in these exposures but the Roadwater Limestone is not present here, although it is present in old quarries a few score metres to the east and west. Locally this limestone is known to increase in thickness and in some instances it does so for reasons of tectonics rather than deposition. At Wheddon Cross its absence may be due to reasons of either kind. The local quartz-veining suggests that strike-faulting may be responsible.

4 **Honeywell** (ST 916 374) About 2 km south of Wheddon Cross the A 396 enters a narrow steep-sided valley in which there are numerous small and rather overgrown roadside exposures. Care must be taken to avoid traffic in this valley as there is only a narrow and discontinuous verge beside the winding road. At Honeywell the old quarry beside the road shows monotonous grey unfossiliferous slates, referred to in this region as Sticklepath Slate. Fossils elsewhere in this unit include brachiopods and crinoids and the formation is correlated with the lower part of the Morte Slates. In this valley the beds dip southwards at about 50° and are largely coincident with the cleavage.

5 **Quarme** (SS 918 363) Further exposures on the eastern side of the road occur between the last locality and this old quarry which is on the east side of the minor road to Quarme Hill at its junction with A 396. The quarry is cut into well-bedded light green sandstones and minor laminated siltstones of the Brendon Hill Beds. Some of the sandstone bands are micaceous and, because the bedding and cleavage largely coincide (at about 50° to the south), poorly preserved plant remains may be found. The Brendon Hill Beds may represent the mid-part of the Morte Slates.

6 **Kersham** (ST 945 389) The steep and narrow lane from the Quarme Valley past North Quarme Farm should be followed to Quarme Hill and the road to Timberscombe taken as far as Couple Cross (SS 951 387). Here a trackway leads west down the hill to Kersham. The old quarry is in a field on the north side of the track about 0.6 km from the road junction. The Roadwater Limestone is exposed and fossils from here have included *'Cystiphyllum' secundum*?, *Phillipsastraea* sp. and stromatoporoids.

7 **Slowley Wood** (SS 986 380) From Couple Cross the road to Luxborough and Roadwater passing the old quarries in Roadwater Limestone just north of Nurcott Farm, descends eastwards and continues on to Kingsbridge. Beyond Kingsbridge it enters Druid's Combe with Slowley Wood to the north. Leaving Kingsbridge turn left on the Rodburgh road and immediately left again up a track; after 100 m the quarry is seen on the right behind the Brendon Forest cabin. In this large old quarry occur lenticularly bedded, massive, buff brown and pinkish brown quartzitic sandstones with

minor slate interbeds and lenses of intraclast conglomerate. The rocks dip southwards at about 25° and are clearly recognisable as Hangman Grits.

The narrow valley (Perley Combe) to the west of Slowley Wood may in part be developed in response to the presence of the N–S striking Kingsbridge Fault. The effect of the fault has been to throw the Avill Group against the underlying Hangman Grits on the south side of Croydon Hill. It has, presumably, also been responsible for the rather sharp deflection northwards of the course of the stream at Kingsbridge.

8 Druid's Combe (SS 993 375) Druid's Combe runs eastwards along the strike of the rocks about the junction of the Hangman Grits and the Avill Group. Roadside exposures are common on the southern side of the combe and many are in the grey brown slates and the siltstones at the base of the Avill Group.

9 Treborough Quarry (ST 015 367) The road through Druid's Combe leads to Roadwater where it is joined by the road from Treborough and Brendon Hill. Some 2 km south of the junction the Treborough road passes a very large and deep old quarry on its southern side (entrance on east side off main road). This pit was originally sunk for the Roadwater Limestone which was extracted from the quarry along a tunnel under the road. The present excavation also reveals the Cutcombe Slate and the Leighland Beds. The quarry is now used as a West Somerset District Council dump. The limestone can be seen as a pinkish grey band about two-thirds of the way up the quarry face opposite the road. The slates comprise about 18 m of light grey material with narrow transitional bands into the limestone. The limestone is cleaved, crystalline and fossiliferous and is only about 2 m thick, less than a quarter of the thickness at Roadwater 2 km to the NE.

10 Roadwater (ST 034 384) The exposure of Roadwater Limestone in Harper's Wood at the northern end of Roadwater Village is in a large disused and very overgrown quarry. The quarry is invisible from the road and difficult to find. In the village cross the Recreation Ground toward the river; leave by the stile and proceed over footbridge northwards along edge of the wood; at the end of the wood turn back and climb diagonally uphill past the old kilns. The quarry can be found east of the kilns. It consists of some 10 m or more of pinkish-grey crystalline limestone and overlying pinkish cleaved limestone. Fossils recorded from this locality include *Heliophyllum halli*, *Thamnophyllum caespitosum*, *Thamnopora reticulata*, *Thamnopora* sp., bryozoa and crinoid ossicles. Other exposures nearby have yielded brachiopods, gastropods and bryozoa. The dip of the beds here is ENE, marking the eastern closure of the Croydon Anticline.

3
The sedimentology and structure of the Upper Palaeozoic rocks at Portishead

B. P. J. Williams and P. L. Hancock

Objectives

The Upper Palaeozoic rocks of the Portishead area examined in this excursion outcrop from Portishead Pier in the east, around Portishead Point (on some maps named Battery Point) and south-westwards to Kilkenny Bay. Rocks ascribed to the Old Red Sandstone facies of the Devonian, and the Tournaisian and Westphalian stages of the Carboniferous are well exposed in these coastal sections. A discontinuous cover of Triassic Dolomitic Conglomerate rests unconformably on the folded and faulted Upper Palaeozoic rocks.

The purpose of this excursion is to examine those features of the coastal outcrops which help in evaluating the sedimentary and tectonic history of the Upper Palaeozoic rocks. Variations in lithofacies, particularly in vertical sections, together with changes in tectonic style affecting these Upper Palaeozoic rocks are outlined.

Special Features

All the sections described are included on the IGS 1:25 000 Sheet ST 47. Access to the sections is easy and all the sequences, with the exception of the Pennant Sandstone, can be studied at any stage of the tide. The rocks in the centre of Woodhill Bay, near the sea wall, are obscured by Recent tidal flat sediments and estuarine alluvium.

Maps

Ordnance Survey	One-Inch Sheet 155 (Bristol & Newport)
	1:50 000 Sheet 172 (Bristol & Bath)
Geological Survey	One-Inch Bristol District
	1:25 000 Sheet ST 47 (Clevedon & Portishead)

Main References

Butler, Williams & Bradshaw (1972); Pick (1964 a & b); Stead & Williams (1973).

Outline Geology

The Upper Palaeozoic sequences display a wide variety of rock types (Fig. 1) but the area generally lacks good faunal assemblages. The Pennant Sandstone, of Upper Coal Measure (Westphalian) age, comprises a sequence of alluvial sediments which includes intraformational conglomerates and thick cross-bedded lithic sandstones with infrequent mudstone horizons. The Lower Carboniferous Black Rock and Lower Limestone Shale Groups are dominated by dolomitized limestones which are tectonically deformed and in places reddened by staining from the Triassic cover. Much of the original shell content of these shallow marine rocks has been destroyed by dolomitization. The Old Red Sandstone facies is made up of a thick sequence of alluvial sediments exhibiting mainly coarse clastic lithotypes including extraformational pebble conglomerates and lithic sandstones. The finer

SYSTEM		STAGE	FORMATION	THICKNESS (metres)	GENERAL LITHOLOGY
CARBONIFEROUS	UPPER	WESTPHALIAN	PENNANT SANDSTONE	18–20	Coarse to medium-grained, green-brown lithic sandstones displaying much trough cross bedding. Erosion surfaces overlain by intraformational conglomerates comprising clay ironstone pebbles and plant debris. Mudstone horizons are rare.
	LOWER	TOURNAISIAN	BLACK ROCK DOLOMITE	155*	Fine-grained, grey dolomitic limestones with chert horizons. Occasional bioclastic limestones with shelly faunas.
			LOWER LIMESTONE SHALE	100*	Alternating lenticular limestones and calcareous siltstones. Limestones contain skeletal debris and well preserved fossils whereas siltstones are intensely bioturbated. Dolomitized and red stained throughout.
DEVONIAN	UPPER OLD RED SANDSTONE	FARLOVIAN	PORTISHEAD BEDS	194	Variable sequence of red and green coarse pebbly sandstones, coarse to fine-grained sandstones and siltstones. Fossil fish locally abundant. Prominent green sandstone unit in middle of sequence. Fining-upward sequences present with much cross bedding and associated depositional sedimentary structures.
			WOODHILL BAY CONGLOMERATE	4	Polymictic roundstone conglomerate of extraformational origin.
	LOWER OLD RED SANDSTONE	?BRECONIAN	BLACK NORE SANDSTONE	292 (min.)	Dominantly red, fine to medium grained quartzitic sandstones internally flat and cross bedded. Intra-formational conglomerates common, siltstones rare. Many pedogenic limestone horizons in the sequence.

Fig. 1 Stratigraphy of the Upper Palaeozoic rocks of Portishead. *Thickness taken from IGS Sheet ST 47 for Portishead area as a whole. Other thicknesses represent field measurements.

mudstone units only occur in abundance in the upper part of the sequence. Coarse sharpstone conglomerate beds are abundant as the Triassic cover rocks but, in places, the sandy facies of these Keuper deposits are present.

The outcrop distribution of these Upper Palaeozoic rocks is largely controlled by an ENE–WSW trending anticline which produces a prominent ridge extending from Clevedon in the SW to Portishead in the NE, a distance of some 5 km. Lower Old Red Sandstone and Lower Carboniferous rocks dip at moderate angles on the SE limb of the fold. The NW limb of the fold is not exposed as it occurs beneath the Severn Estuary. The fold plunges WSW at its southern end and ENE at its northern end. Thus, the oldest rocks in the area, the Lower Old Red Sandstone, are present only between Kilkenny Bay and Charlcombe Bay (ST 432 751). Between Portishead Point and Portishead Pier (the Eastwood ridge of Reynolds & Greenly 1924) at the northern end of the area described, the strike is E–W rather than ENE and the Upper Palaeozoic rocks are bounded to the south by east to west strike faults downthrowing to the north (Reynolds 1921).

The structure of the Portishead region was first described in detail by Reynolds & Greenly (1924) who had earlier (1923) outlined the stratigraphy of the Upper Palaeozoic rocks of the area. The stratigraphy and sediments of the Old Red Sandstone were discussed by Wallis (1927a) and Pick (1964a,b). The remapping of the Bristol area by officers of the Institute of Geological Sciences resulted in the publication in 1968 of a special sheet (ST 47) of the Clevedon-Portishead district on a scale of 1:25 000. The results of recent work on the sedimentology of the Old Red Sandstone – Lower Limestone Shale transition and the Pennant Sandstone are contained in Butler, Williams & Bradshaw (1972) and Stead & Williams (1973). Stratigraphical problems still exist in the sequences at Portishead. For example, the exact position of the Pennant Sandstone succession at Portishead within the Upper Coal Measures is not known with certainty. The relationship of the Woodhill Bay Conglomerate to the Portishead Beds and Black Nore Sandstone requires further investigation and the age of the latter formation has still to be accurately fixed (Kellaway & Welch 1966,p. 5).

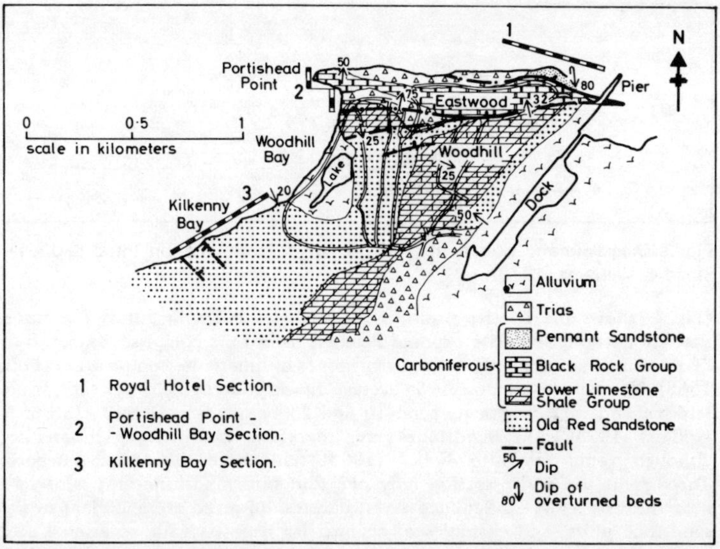

Fig. 2 Geology of the Portishead area showing location of sections described in the text. Geology slightly modified after IGS 1:25 000 Sheet ST 47.

Itinerary

The itinerary commences in the Pennant Sandstone at the Royal Hotel section. Other localities are described in descending stratigraphical order and in an anticlockwise geographical sense. All the localities can be comfortably traversed in one day working from the Royal Hotel to the steps in Kilkenny Bay near the top of the Black Nore Sandstone. The section numbers correspond to those shown in Fig. 2. The circled letters indicate the individual localities within the sections and are noted on the relevant figures; in text they appear in parenthesis thus (A).

1 **Royal Hotel Section** From Portishead town centre take South Avenue and South Road to the junction with Leigh View Road. Turn right to park near the Royal Hotel. Steps nearby descend to the pebble beach to commence the section through the Pennant Sandstone; this small outcrop is the only coastal section in these rocks east of Dyfed. The Pennant Sandstone here is isolated from the main development of Upper Carboniferous rocks of the Nailsea Basin and Clapton to East Clevedon area by folding and faulting. At this locality the Pennant Sandstone is bounded to the south by a northerly-downthrowing strike fault. This fault is probably not of any great magnitude since in this region the Upper Carboniferous unconformably overlies the Lower Limestone Shale Group of the Lower Carboniferous (Kellaway & Welch 1948, p. 27).

The exact stratigraphical level of the Portishead Pennant Sandstone within the Upper Carboniferous is not known with any certainty because of the absence of coal seams to act as marker horizons. On lithofacies evidence the rocks appear to be either part of the Mangotsfield Group or the upper part of the Downend Group (lower division of the Upper Coal Measures).

The Pennant Sandstone of the Royal Hotel section crops out from ST 4726 7767 to ST 4761 7752 comprising essentially a strike section with beds dipping mainly towards the NNE. Some overturning of beds is present at the eastern end of the section and is outlined below. The section described is from east to west (Fig. 3) and a vertical log of the sedimentary sequence

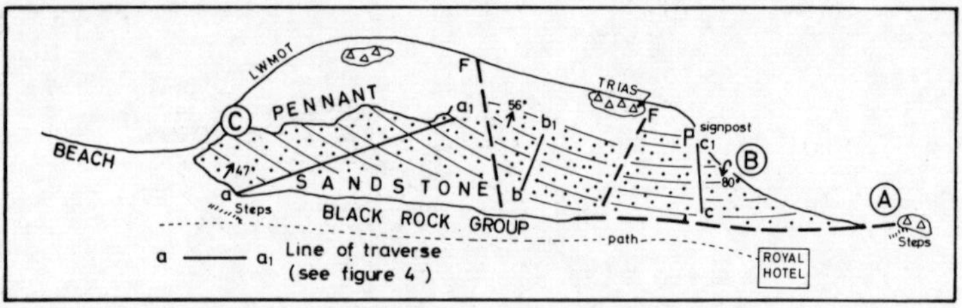

Fig. 3 Diagrammatic sketch map of the Royal Hotel Section (modified after Stead & Williams 1973).

(Fig. 4) shows the relative position of lithologies and sedimentary structures that are encountered. The exposed Pennant Sandstone comprises some 18 to 20 m of medium to coarse grained sandstones of litharenite composition (Folk 1968). The sandstones possess an average quartz content of 57 per cent, while feldspar and rock fragments total 10 and 33 per cent respectively (Stead & Williams 1973). The sandstones are, therefore, classified as litharenites although some fall into Folks' (1968) feldspathic litharenite category. Throughout the entire section only one thin silty mudstone unit is present near the base (Fig. 4). Sedimentary structures, however, are abundant in the sandstone units, and examples of channel lag deposits with occasional coal conglomerates are readily seen. In places reddening of these otherwise drab green and brown sandstones is attributed to hematite staining through downward migration of fluids from the Triassic cover.

At ST 4761 7752 at beach level (A) the complex geology of the extreme eastern end of the Royal Hotel section can be studied. Here the Black Rock Dolomite is faulted against the Pennant Sandstone, the north-dipping normal strike fault being easily located in the low reefs of the foreshore. Low reefs of Black Rock Dolomite are present near the pier where they dip south at moderate to high angles.

At the foot of the steps the Triassic Dolomitic Conglomerate directly overlies the Black Rock Dolomite in a small cliff. Some 2 m of sharpstone conglomerate is present in this section and these rocks dip in a northerly direction at a low angle. Large, angular clasts, up to 45 cm long axis dimension and mainly of limestone, are enclosed in a yellow-brown matrix of calcareous sandstone. In calcite-filled vugs within the Conglomerate small patches of galena and azurite may be seen.

The Black Rock Dolomite at this locality is reddened by staining from the overlying Dolomitic Conglomerate. Bedding in the Black Rock Dolomite is obscure and the rocks are disrupted in places due to the proximity of the fault. Many infilled fissures containing red calcite are seen within the Black Rock Dolomite.

A further point of interest at this particular stop is the presence, in fair abundance, on the gravel beach of angular blocks of biotite-garnet gneiss. Obviously exotic to this area two suggestions as to their occurence have been put forward: the boulders may be of glacial origin, erratics produced by southerly moving ice during the Quaternary; the material may simply represent discarded ship's ballast. A point in favour of the first theory is the presence of a 50 ton glacial erratic of identical mineralogical composition at Porthleven in Cornwall. Against this idea is the fact that the gneiss is totally restricted to this one locality along the Portishead coast and that this locality is in juxtaposition to Portishead Docks. Their occurrence, therefore, is a matter for further debate.

The outcrops at (B) (ST 4753 7754) are entirely within the Pennant

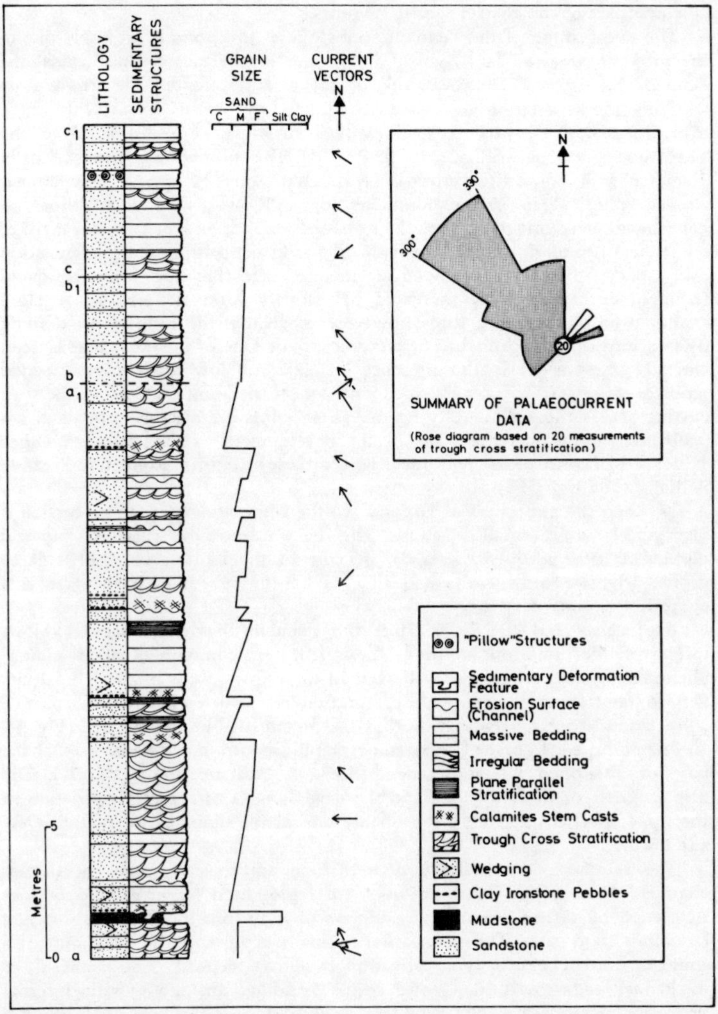

Fig. 4 Vertical profile of the Portishead Pennant Sandstone. (Line of traverse a–a_1 is continuous, whereas b–b_1 and c–c_1 may or may not be continuous.) Modified after Stead & Williams (1973).

Sandstone and are close to traverse C–C_1, near the warning sign (Fig. 3). A vertical log of this part of the section is shown and labelled C–C_1 on Fig. 4. The Pennant Sandstone at this stop is locally overturned and dips at about 80° south. The rocks are coarse to medium grained sandstones containing discrete horizons of granule conglomerate with coal clasts. In other places coal, clay ironstone and mudstone make up a lag deposit at the base of the coarser sandstone units. Internally the sandstones are frequently trough cross-bedded with sets up to 1 m thickness. Erosion of medium grained micaceous sandstones with primary current lineation is present beneath many of the channel bases of the coarser sandstone units.

The overturning of the Pennant Sandstone at this point is probably due to the proximity of the fault which throws the Black Rock Group against the Pennant Sandstone. The 'way up' of the strata is readily determined by studying the truncation of trough cross bedding at this locality. In these overturned beds spectacular 'pillow-like' structures are displayed in the Sandstone (Stead & Williams 1973, P1.VII). The 'pillows' are almost 1 m in diameter with concentric internal layers. With one exception they appear almost spherical in the exposed section. 'Pillows', similar to those at Portishead, were originally thought to have formed in one of two ways; either by vertical foundering of sediment shortly after deposition or by aggregation into concretions due to chemical changes affecting the unconsolidated sediment as a result of increased pH shortly after deposition. A close examination of these structures, however, reveals that the 'pillows' are defined by concentric joints probably of tectonic origin. One of the most readily seen clues to this origin is the presence of adjacent low angle cross bedded sandstone units which display incipient concentric joints which can be seen cutting across the sedimentary layers. These joints are discontinuous and are probably an intermediate stage in the development of the 'pillows'. Other evidence in favour of the tectonic origin of these structures is given in Stead & Williams (1973, p. 312).

Between the exposures at (B) and (C) the Pennant Sandstone succession is disrupted by a few small dip faults. The rocks are now the right way up and dipping at 56° north-east (Fig. 3). Because of the faulting it is difficult to decide whether or not the sequence is continuous and that Fig. 4 is a composite sedimentary log.

At locality (C) (ST 4730 7762) the Pennant Sandstone again exhibits interesting deformation features. These features, commonly called 'slump' phenomena, are particularly well seen in these lower beds in the Sandstone. Penecontemporaneous sediment deformation has occurred in the lower part of a thick sandstone unit which overlies the 30 cm silty mudstone bed (Fig. 4). At one point, a load cast of an asymmetric pillow form may be observed at the base of a slipped sandstone sheet (Stead & Williams 1973, P1.VII). The displacement of the mudstone and the development of rotated lamination in the upper part of the unit are probably a result of sliding towards the WNW and north.

The sequence above the silty mudstone unit reveals some spectacular channel phenomena. These features are represented by erosional surfaces succeeded by coarse grained sandstones with clay ironstone lag deposits and *Calamites* stem casts. This association normally represents the basal unit of a fining-upward sequence, typical of modern alluvial deposits, and is a result of in-channel sedimentation. Trough cross bedding, analogous with internal bedforms of modern alluvial sand bars, is ubiquitous in this part of the section (Fig. 4) and the orientation of these sedimentary features reveals the direction of flow at the time of deposition. In the Portishead Pennant Sandstone the main mode is directed towards the NW but some vectors are at variance with this mode and these are ascribed to meandering effects (Fig. 4). The implication is that the main direction of river flow was towards the north-west. This direction conforms to the regional palaeocurrent patterns in the Lower Pennant sequences of Bristol and Somerset, South Wales and the Forest of Dean (Kelling 1968, Gayer & Stead 1971).

Westwards beyond this point outcrops of Pennant Sandstone become discontinuous and the formation crops out only in a few small inliers through the Triassic cover. For example at ST 4715 7762 nearly vertical Pennant Sandstone can be seen beneath a cover of nearly horizontal Triassic rocks. A little further to the west (ST 4687 7756) low reefs and cliff exposures of the

Triassic rocks allow detailed study of these sediments. Here the Trias is yellow and red in colour and locally comprises coarse to fine grained sandstones which are generally flat bedded. Conglomeratic horizons are still present and may contain large clasts of limestone. Some channel structures may be observed in these Triassic rocks and the unconformable relationship with the Pennant Sandstone is well displayed.

2 **Portishead Point – Woodhill Bay Section** This section through part of the Lower Carboniferous succession is examined from north to south. Exposure is near-continuous through the Black Rock Dolomite and Lower Limestone Shale Group from a point 100 m ENE of Portishead Point to the northern end of the sea wall in Woodhill Bay. The sedimentary rocks in these outcrops are varied but the outstanding geological feature of this section is its structure (Fig. 5). In order to appreciate the folds exposed on Portishead Point the traverse should be commenced about 10 m ENE of the Beacon on the Point. To reach the Beacon, ascend the steps past the swimming pool, cross the grass, and then descend onto the low cliffs. Here the most northerly folds to be examined are exposed.

The general lithological features of the Black Rock Dolomite (A) are described before the traverse to examine the structural details is commenced. The Black Rock Dolomite appears from beneath the cover of Triassic Dolomitic Conglomerate some 100 m ENE of Portishead Beacon at ST 4648 7760. This sequence of carbonate rocks mainly comprises fine grained dolomitic limestones with chert horizons. Although mapped as Black Rock Dolomite the limestones are not dolomitized throughout their entire thickness. In places, particularly beneath the catwalk leading to the Beacon, bioclastic limestones are to be found. Here locally abundant patches of limestone are rich in fossils and include species of gastropods, crinoids and brachiopods, particularly spiriferids. Much of the original shell content of the limestones has been destroyed by dolomitization although analyses of the Black Rock Dolomite show that its magnesia (MgO) content is rarely in excess of 15 per cent (Kellaway & Welch 1955). In addition many of the shells in the Dolomite are replaced by chert. At this locality it also appears that the lower Black Rock Group contains more finer grained micritic limestones than equivalent horizons in eastern Mendip. It has been suggested (Butler 1972) that this part of the succession of shallow marine rocks at Portishead represents carbonate build-up on a shallow platform. Development of skeletal sand shoals on the outer part of the platform would have provided sheltered inner reaches where these finer grained limestones accumulated.

IGS 1:25 000 sheet ST 47 shows the Black Rock Dolomite at Portishead Point as dipping north at about 40°. However, along a section immediately north and east of the approximately N–S concrete pathway which passes beneath the catwalk leading to the Beacon several small folds can be seen (Fig. 5(a)). Viewed from the west the folds north of the catwalk at the start of the traverse are Z-shaped in profile. These asymmetric, inequant folds plunge gently east and possess axial planes inclined steeply to the north. Along the southern side of the point, which is reached by following the concrete pathway beneath the catwalk to a point level with the ruined watchtower, the otherwise uniformly steep dip to the north is locally interrupted by several markedly inequant folds which are S-shaped in profile, and which plunge east at angles up to about 25°. The axial planes of these folds also dip to the north. The relatively steep plunge means that the marked V-shaped outcrops of the synforms on the narrow wave-cut platform at the base of the cliff open to the east. The hinge zones of some of the folds are relatively tight and show thickening of some of the beds in fold cores.

The presence of Z-shaped, south verging parasitic folds north of the catwalk, and S-shaped, south verging parasitic folds along the southern side of

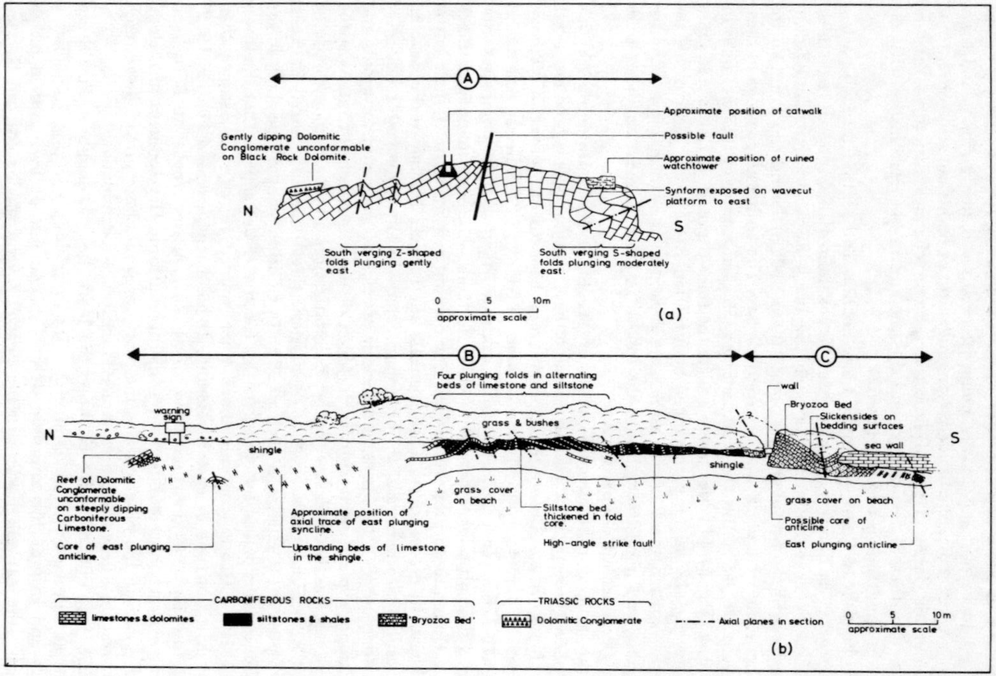

(a) Composite schematic cross section of structures in the Black Rock Dolomite along an approximately N–S line east and north of the concrete pathway beneath the catwalk to Portishead beacon.

(b) Sketch cross section of structures in the Lower Limestone Shale in the low cliff and foreshore at Woodhill Bay. Not all limestone and siltstone beds are shown individually.

A, B and C indicate the localities referred to in the text.

Fig. 5 Tectonic structures in Carboniferous rocks at Portishead Point and Woodhill Bay.

the Point suggests that the Black Rock Dolomite of Portishead Point is involved in a south verging anticlinorium with a normal northern limb and an overturned southern limb (Fig. 5(a)). The exact position of the axial trace of the anticlinorium, which may be faulted, is difficult to detect, but it probably lies a few metres south of the catwalk.

The folds on Portishead Point differ from those in the Lower Limestone Shale of Woodhill Bay in two important respects. Firstly, their profile shapes differ, perhaps as a consequence of the contrasting lithological assemblages involved in the two areas, and secondly they differ in their direction of vergence. The folds at Portishead Point verge south while the folds in the Lower Limestone Shale verge north. The E–W axial trend of both groups of folds is at variance with the WSW–ENE trend of the Clevedon-Portishead anticline which is the principal structure entering the Portishead district.

Immediately south of the promontory of Portishead Point a shallow valley (B) (ST 4646 7748) infilled with Triassic sediments (IGS Sheet ST 47) obscures the relationship between the Black Rock Dolomite and the Lower Limestone Shale.

The exposed sequence of rocks in this section extends from the red warning sign by the swimming pool (ST 4646 7746) to a point close to the start of the sea wall (ST 4644 7739). The section is in the Lower Limestone Shale Group and is composed essentially of alternating lenticular limestone units and intensely burrowed siltstones folded into a series of asymmetrical anticlines and synclines. The transition beds from the Old Red Sandstone into the Lower Carboniferous are nowhere seen on the foreshore at Woodhill Bay. However excavations for a new building site inland at Woodhill (ST 4689 7733) temporarily exposed a section across this transitional group (Butler, Williams & Bradshaw 1972). The lithological variations and associated sedimentary structures from this temporary section are summarized in Fig. 6 which may be used for comparison with similar lithologies and sedimentary structures exposed in the foreshore sequence at Woodhill Bay.

In Woodhill Bay (ST 4646 7743) the limestone units are made up of a mixture of highly abraded hematitic skeletal debris and well preserved fossils, including articulated crinoid stems and fairly common examples of the tabulate coral *Vaughania (Cleistopora)* – the index fossil of the Lower Limestone Shale (Matthews, Butler & Sadler 1973). In addition to the lenticular limestone units the interbedded siltstones exhibit a wide variety of burrows. Forms recognized include *Chondrites, Teichichnus, Rhizocorallium* and ?*Thalassinoides.* The Lower Limestone Shale lithologies are strongly affected by Triassic alteration, and the Dolomitic Conglomerate overlies the folds on the beach. This weathering has brought about the almost complete dolomitization of the limestones and has produced a red colouration by redistribution of iron oxides.

The Lower Limestone Shale sediments at Portishead are thought by Butler (1972) to have accumulated under intertidal and subtidal conditions in which algal activity is restricted and the skeletal debris comprises crinoid and brachiopod material. Detrital phosphate in the sequence (see Units 3 and 4, Fig. 6) indicates lag deposits with an increase in shell debris. Influxes of hematitic skeletal debris, frequently associated with poor sorting and mega-ripples, suggest that these deposits were brought into the area of sediment accumulation by exceptional currents (Butler 1972). Higher in the sequence hematitic horizons become finer in grain size and sorting increases. It appears that this debris was washed around in the tidal flat and finally concentrated in a sediment bank. It is above this bank in the Woodhill Bay section that the tabulate corals and crinoids become abundant. According to Butler (1972) it is difficult to estimate the depth zones (Seilacher 1967) involved from the trace fossil assemblages (burrows) found in this sequence.

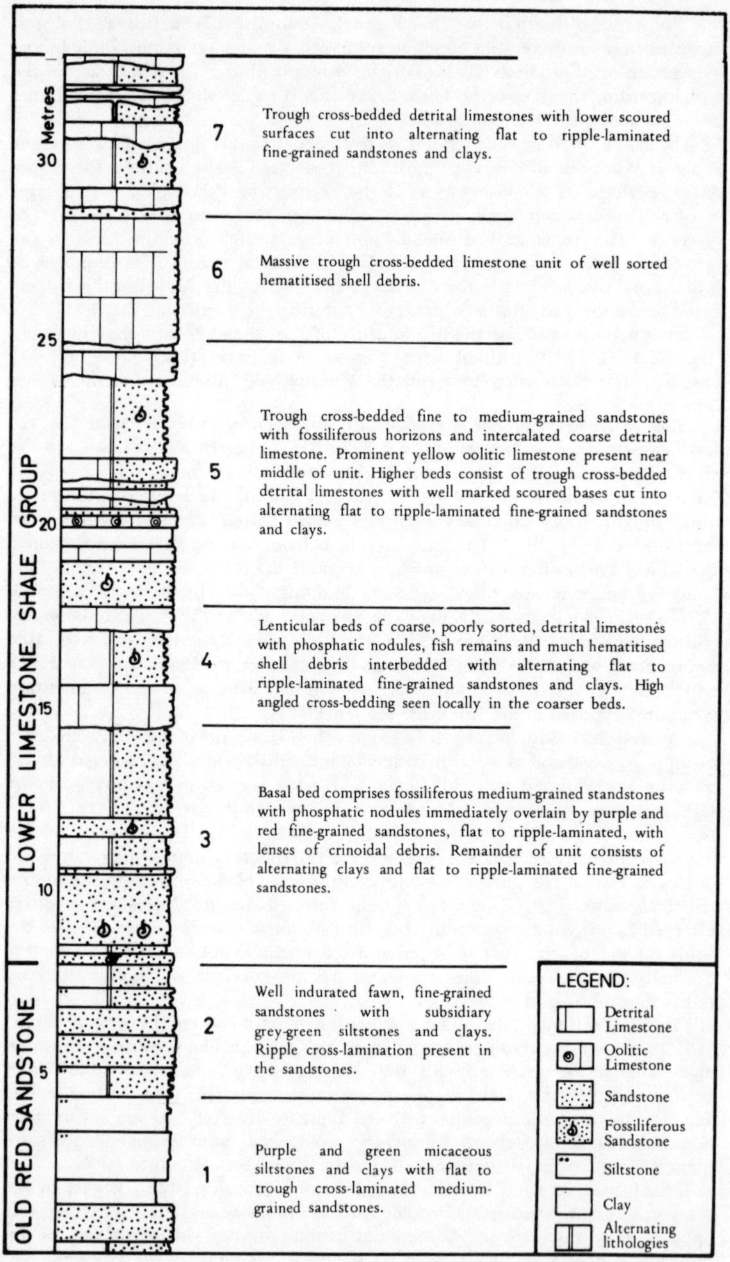
Metres
30
25
20
15
10
5
LOWER LIMESTONE SHALE GROUP
OLD RED SANDSTONE
7
Trough cross-bedded detrital limestones with lower scoured surfaces cut into alternating flat to ripple-laminated fine-grained sandstones and clays.
6
Massive trough cross-bedded limestone unit of well sorted hematitised shell debris.
5
Trough cross-bedded fine to medium-grained sandstones with fossiliferous horizons and intercalated coarse detrital limestone. Prominent yellow oolitic limestone present near middle of unit. Higher beds consist of trough cross-bedded detrital limestones with well marked scoured bases cut into alternating flat to ripple-laminated fine-grained sandstones and clays.
4
Lenticular beds of coarse, poorly sorted, detrital limestones with phosphatic nodules, fish remains and much hematitised shell debris interbedded with alternating flat to ripple-laminated fine-grained sandstones and clays. High angled cross-bedding seen locally in the coarser beds.
3
Basal bed comprises fossiliferous medium-grained standstones with phosphatic nodules immediately overlain by purple and red fine-grained sandstones, flat to ripple-laminated, with lenses of crinoidal debris. Remainder of unit consists of alternating clays and flat to ripple-laminated fine-grained sandstones.
2
Well indurated fawn, fine-grained sandstones with subsidiary grey-green siltstones and clays. Ripple cross-lamination present in the sandstones.
1
Purple and green micaceous siltstones and clays with flat to trough cross-laminated medium-grained sandstones.
LEGEND:
Detrital Limestone
Oolitic Limestone
Sandstone
Fossiliferous Sandstone
Siltstone
Clay
Alternating lithologies

Fig. 6 Composite stratigraphical section of the Old Red Sandstone – Lower Limestone Shale transition at Woodhill, Portishead. (after Butler, Williams & Bradshaw, 1972).

The rocks of the Lower Limestone Shale at localities (B) and (C) are deformed by at least 10 adjacent small buckle folds (Fig. 5(b)). Relatively large numbers of small folds are not uncommon in deformed multilayer sequences.

Some of the inequant asymmetric folds possess relatively short, steep, or slightly overturned north dipping limbs and longer moderately inclined south dipping limbs. The axial planes of the folds dip south at angles between about 55° and 85° Both the direction of the dip of the axial planes and the relative limb lengths indicate that the folds verge north. Viewed from the west some fold profiles define S-shapes which suggests that they lie on the moderately inclined, south dipping, normal limb of a larger fold. Direct observation of fold hinge lines shows that they mainly trend a few degrees south of west, and plunge gently either to the east or the west. A few hinge lines are noticeably curved in plan and exhibit culminations and depressions in profile. Fold axes, which have been statistically defined from stereographic π diagrams, are orientated similarly to the hinge lines but the two directions are not necessarily exactly parallel for individual folds. Fold style is mainly parallel or flattened parallel. The siltstone interbeds between limestones commonly exhibit the effects of accommodation in the cores of some of the tighter folds.

In a N–S traverse of the foreshore at locality (B) the following folds display features worthy of detailed examination (Fig. 5(b)). At the first exposure of the solid geology south of Portishead Point about 1 m of Dolomitic Conglomerate dipping moderately north rests unconformably on nearly vertical limestones belonging to the Lower Limestone Shale. For a few metres south of here shingle largely obscures the structure until the hinge zone of small easterly plunging anticline in limestone is reached. South of this hinge zone, which is well exposed in three dimensions, the position of the axial trace of a syncline can be determined by inspecting the V-shaped outcrop formed by its limbs exposed in the hard limestone bands which form a slightly raised relief on the foreshore. The V of the outcrops opens to the east, indicating the direction of plunge of the syncline.

Immediately south of this part of the section four, adjacent, asymmetrical gently plunging anticlines and synclines in limestones and interbedded siltstones are well-exposed in the low cliff and on the beach. Some of the short, steep limbs of these northerly verging folds are overturned by a few degrees. In the cores of some of the folds the effects of flow and accommodation in the siltstones are visible. Siltstone beds are generally thicker in the hinge zone of the fold than they are on the limbs. There is a less noticeable difference in the thickness of limestone units between hinge zones and limbs. Culminations and depressions along some of the curving hinge lines of the gently plunging impersistent folds can be seen on the foreshore.

For about 15 m south of the southernmost syncline of the above group of four folds exposure is restricted and comprises mainly soft, weathered calcareous shale and thin calcarenite bands. Nevertheless it is possible to inspect in the cliff an example of a high-angle strike fault separating a limestone from steeply dipping siltstones and calcarenites. The slickenside lineations on the fault plane pitch at a high angle to the strike of the fault indicating that there has been dip-slip movement on the fault.

The southernmost locality (C) in the Portishead Point – Woodhill Bay section is found immediately adjacent to the northern end of the sea wall (ST 4644 7738). Here a massive bed consisting of well sorted carbonate debris in a complex of lenticular trough cross-bedded units and including abundant hematized crinoid ossicles is seen. This bed was referred to by Reynolds & Greenly (1923) as the 'Bryozoa Bed' and the foreshore outcrop is directly comparable to a thick cross bedded unit (Unit 6) present in the Woodhill road cutting (Fig. 6). Because of the relative proximity of these two exposures – a distance of some 300 m - it seems probable that they represent the same bank-like sediment accumulation. Within this foreshore example at Woodhill Bay burrows of a *Thalassinoides*-type trace fossil can be examined.

The 'Bryozoa Bed' is involved in two anticlines and an intervening syncline all of which verge north. The core of the northern fold, an overturned anticline, occupies a marked gulley in the cliff. The syncline to the south plunges gently west and displays on some bedding surfaces slickenside lineations orientated approximately perpendicular to its hinge line. The slickensides indicate that the mechanism of fold development involved some flexural-slip. The arrangement of the lenticular trough cross beds shows that both limbs of the fold are normal, that is they have not been overturned. The anticline south of the syncline is the last exposure of the solid geology on the foreshore before the shingle and alluvium of the centre of Woodhill Bay. The fold plunges gently east in contrast to the westerly plunge of the adjacent syncline to the north.

3 **Kilkenny Bay Section** An excellent sequence of Upper and Lower Old Red Sandstone is exposed in clean cliff sections from the southern end of the Woodhill Bay sea wall through Kilkenny Bay and beyond Black Nore Point. The section has been eroded through the core of the northerly plunging part of the Portishead anticline, thus the oldest rocks – those of the Lower Old Red Sandstone – appear at the southern end of Kilkenny Bay and extend beyond Black Nore Point towards Charlcombe Bay. Along this oblique strike section the lowest units of the Old Red Sandstone rocks are not exposed. The Upper and Lower Old Red Sandstone successions were named the Portishead Beds and Black Nore Sandstone respectively by IGS officers when remapping the area (Kellaway & Welch 1948, 1955). Throughout the section the rocks generally dip at low angles to the SE. A discontinuous cover of Triassic Dolomitic Conglomerate blankets the Old Red Sandstone strata and, in places, excellent examples of angular unconformity are present.

The Old Red Sandstone of Portishead, in the area of its greatest development between Kilkenny Bay and Black Nore Point (ST 443 765), attains a thickness of 486 m. Neither the upper or the lower boundaries of the Old Red Sandstone are found within the section and Pick (1964) suggests that the total thickness of the Old Red Sandstone may be at least 900–1200 m. However the uniformly sandy facies of the Black Nore Sandstone within the Portishead coastal section displays a minimum thickness of 292 m while the Upper Old Red Sandstone (Portishead Beds), of more variable lithology, is 194 m thick, 62 m of which is exposed in Kilkenny Bay (Fig. 7).

The Old Red Sandstone age of these rocks has been recognized since 1862 and vertebrate remains indicating this age have been collected since that time (Baily 1864; Martyn 1876). Many papers have since been published on the stratigraphy of these rocks but Wallis (1927a, b) made a major contribution on the stratigraphy and sedimentology of the Old Red Sandstone of the Bristol district which included a thorough examination of the Portishead section. The most recent and detailed study of the Old Red Sandstone of this area was made by Pick (1964 a, b) who subdivided the succession into twelve 'formations'. These 'formations' were designated, in descending stratigraphical order, A–I and J_1 to J_3, the former comprising the Portishead Beds and the latter the Black Nore Sandstone. The majority of the 'formations' are only seen in the vertical section here described and it is unlikely that any of them, with perhaps one or two exceptions, can be traced laterally, for any distance. Most of Pick's 'formations' simply reflect subtle changes in sedimentary facies with time. For example his 'formations' D and E are part of one major fining-upward cycle in this sequence of alluvial sediments, E being the result of in-channel sedimentation and D comprising the overbank deposits. In this sedimentological setting the units will have very little lateral continuity.

An exception is unit I at the base of the Portishead Beds – the Woodhill

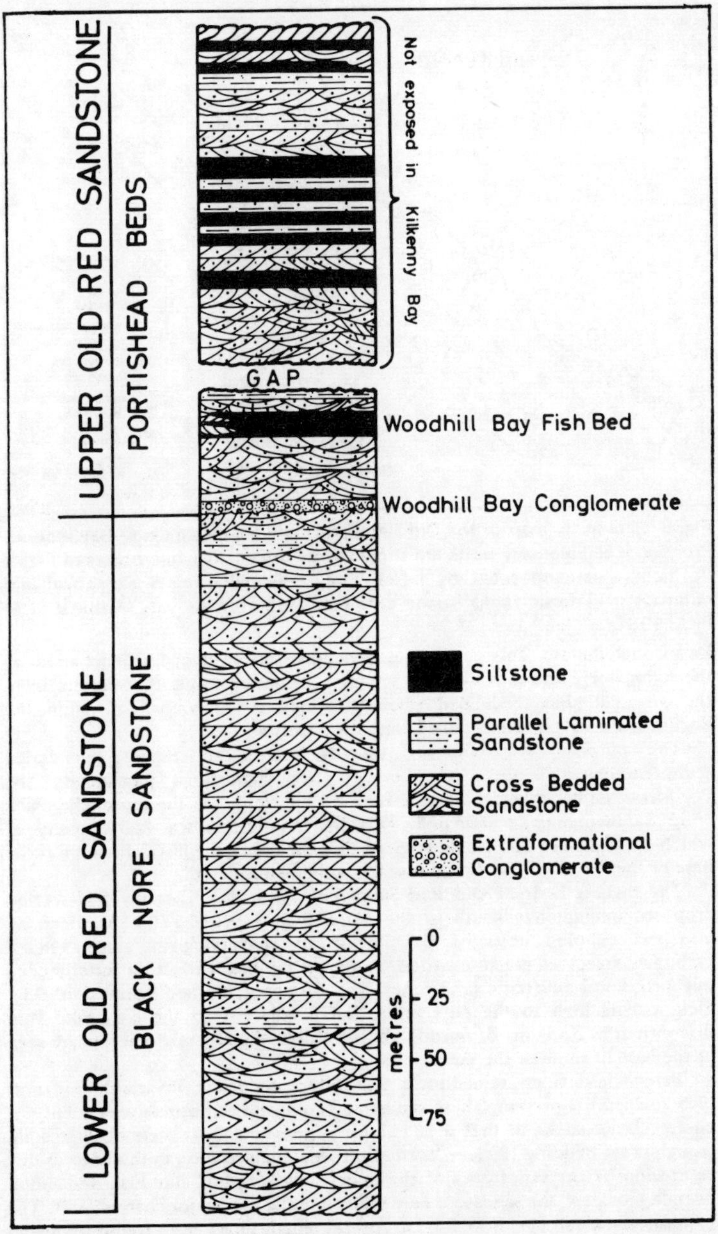

Fig. 7 Generalized vertical profile of the Portishead Old Red Sandstone. Thickness of the Lower Old Red Sandstone based on sections between Black Nore Point and Kilkenny Bay and of the Upper Old Red Sandstone in Kilkenny Bay and Ladye Bay. Modified after Pick, 1964a.

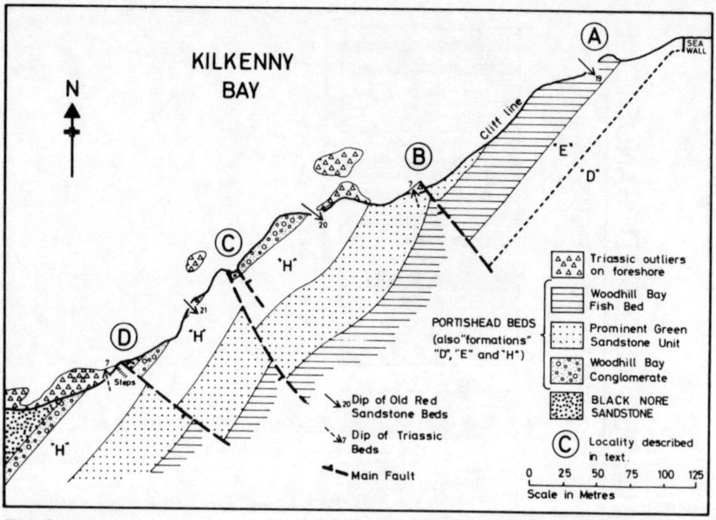

Fig. 8 Geological map of the Old Red Sandstone of the Kilkenny Bay section. Prominent sedimentary units are shown with ornament while others of Pick's 'formations' are indicated by letter. Only the major faults are shown and outcrops of Triassic rocks in the cliff face are not included. Modified after Pick, 1964a.

Bay Conglomerate. This coarse clastic deposit can be traced at least as far as the Avon Gorge and thus warrants separate formation status. Another horizon, the Woodhill Bay Fish Bed (Pick's 'formation' F), occuring within the Portishead Beds has also received formation status.

The purpose of examining this section is to inspect the varied sedimentological features of the Old Red Sandstone, particularly the Portishead Beds between Kilkenny Bay and the steps near the top of the Black Nore Sandstone (ST 4570 7673) (Fig. 8). In addition the faults, many of which exhibit post Triassic movement, and the debatable disconformity at the base of the Woodhill Bay Conglomerate will be considered.

The highest beds of Old Red Sandstone age in the Kilkenny Bay section crop out immediately south of the sea wall at ST 4615 7703 (A). Here the youngest exposed units of the Portishead Beds comprise some 3 m of carbonate-free, red-green mottled, sandy siltstones with thin interbedded micaceous and quartzitic sandstones. This sequence, termed 'formation' D by Pick, occurs high in the cliff immediately adjacent to the sea wall. Thin discontinuous horizons of mud-flake (intraformational) conglomerate are seen at the base of some of the sandstones.

Below these finer grained units a sandstone sequence of variable thickness (7.5 to 17 m) is present. These sandstones comprise an impressive sequence of coarse clastic rocks in that they exhibit superb sets and cosets of large scale trough cross bedding (Pick – 'formation' E). The variation in thickness is due to erosion by the sandstones of the underlying Woodhill Bay Fish Bed and in overall geometry the sandstones are seen to occur in a major channel unit. The lowermost sandstones, in which eurypterid remains have been found (Simpson 1951), contain, at beach level, many vein quartz pebbles whose long axes are frequently imbricate parallel to the foresets of the trough cross bedding. The vein quartz pebbles are sub-angular to well rounded in outline, with low sphericity and average long axis dimension of 13 mm.

Conspicuous erosion, in the form of channelling, can be observed at the base of the pebbly sandstones into the finer grained lithologies of the Woodhill Bay Fish Bed and cutting out this formation in a north easterly direction. From this erosional base through the sandstones and the overlying fine sandstone-siltstone alternations, an excellent example of a fining-upward sequence is observed. This reflects the alluvial character of these rocks resulting from in-channel river sedimentation at first and passing up into overbank or floodplain deposits, the latter exhibiting thin persistent sandstones of crevasse-splay type. Reorientation of cross bedding dip directions in this sandstone sequence indicates a general sediment dispersal direction towards the south (Pick 1964a, p. 219).

Immediately SW of this first exposure, at ST 4613 7699, an indentation in the cliff, due to erosion of the less resistant lithologies of the Woodhill Bay Fish Bed, provides an excellent section through this formation (Fig. 8), the upper part of which represents Beds 5–10 of Wallis's section (1927a, b) and has yielded the most important fossils collected from the Portishead Old Red Sandstone. The main fossils comprise remains of the crossopterygian fish *Holoptychius* referring these rocks to the Upper Old Red Sandstone. Scales of this fish can still be collected in fair abundance at this locality.

The Woodhill Bay Fish Bed formation is 10 m thick and consists of red micaceous siltstones and quartzitic sandy siltstones with sandstone horizons increasing in importance towards the top of the formation. Four prominent sandstones are seen within the formation. Beds vary in thickness between 0.5 and 3 m and contain good examples of both planar and trough cross bedding. Ripple lamination and ripple marked surfaces are also present. In addition some flat bedded units display excellent primary current lineation. At the base of the thickest sandstone unit load casts and flow rolls (Pick 1964a, Pl.8B) indicating deformation, with vertical and lateral translation, of water saturated sediment can be seen. The siltstone-sandstone ratio increases towards the base of the formation, a junction that is readily seen because of a very sharp contact with the underlying, conspicuous, thick green sandstone unit.

At ST 4596 7693 the first major fault in the section is intercepted (B). This fault (Fig. 8), which is not shown on IGS Sheet ST 47, introduces Triassic deposits into the section downfaulting these rocks against the Portishead Beds. The geology within 50 m of either side of this fault is diverse and interesting. To the NW of the fault the Woodhill Bay Fish Bed is seen to rest on a very conspicuous thick green sandstone bed. This bed, which is 14 m thick, has a gently undulating upper surface exposed on a prominent ledge. The bed is a fine to medium grained quartzitic sandstone. Although this sandstone unit is cross bedded throughout its thickness, the large scale foresets are relatively indistinct due to the remarkable grain size isotropy within the bed. Rare, small pebbles and granules of vein quartz are seen close to the base of the bed (Pick – 'formation' G) and the cross bedding indicates dispersal towards the SE. The low angle of tectonic dip (20°) of this sandstone to the SE brings it from beach level to occupy the entire height of the cliff adjacent to the fault.

The nearly vertical normal fault, which brings the green sandstone unit of the Portishead Beds against Triassic deposits, has a downthrow to the SW of some 10 m. According to Pick (1964b) the relationship of the rocks on either side of the fault, can only be explained by assuming pre-Triassic faulting which created a scarp, facing SW, against which the Dolomitic Conglomerate was deposited. Renewed post-Triassic movement along the fault line produced a further displacement of the Triassic rocks. At beach level, in the cliff immediately adjacent to the fault the Triassic rocks are in part made up of orange-brown, relatively fine grained calcareous sandstone which Pick (1964b, p. 446) interprets as a basal Triassic lacustrine beach sand upon which the coarse, unsorted alluvial fan deposits of the Dolomitic Conglomerate were laid down.

To the SW of the fault the Triassic rocks, which dip seaward (NNW) at 7°, form a discontinuous cover on the Portishead Beds. Areas of foreshore are covered by outliers of Dolomitic Conglomerate resting on low reefs of Upper Old Red Sandstone. At the base of the outliers and in the low cliffs good examples of unconformity, between the Triassic and Old Red Sandstone rocks, can be seen. In many places the geology may, on first inspection, appear a little confusing due to the Dolomitic Conglomerate filling in valleys and hollows in the Old Red Sandstone topography which has been exhumed through later erosion. At one such exposure a tongue of Triassic conglomerate is present as a veneer in the trough of an Old Red Sandstone hollow which has been exhumed and, at first glance, it appears that the Trias is overlain by the Old Red Sandstone.

In close proximity (ST 4588 7688) to the unconformity the next sedimentary unit (beneath the prominent green sandstone) of the Portishead Beds is seen to crop out. Pick (1964a) named this unit 'formation' H and it directly overlies the Woodhill Bay Conglomerate. This group of rocks comprises 13.5 m cross laminated and cross bedded sandstones, pebbly sandstones, thin siltstones and conglomerates. Sandstones of varying grain size dominate the sequence. Again they are red in colour, carbonate free and contain cross bedding which indicates a sediment dispersal pattern to the SE (Pick 1964a). Repeated channel avulsion by the river which deposited these sediments is demonstrated by the cross cutting channel sandstones in this part of the section. At one point a spectacular channel complex may be seen. This complex is up to 2 m deep and 6 m wide and contains the eroded remnants of six different fillings revealing constant repetition of erosion and deposition along a preferred stream direction.

Proceeding SW down the succession a number of minor faults are seen beyond ST 4582 7685 where the Woodhill Bay Conglomerate (C) first appears in the section. Almost all the faults in the section affect the Triassic Dolomitic Conglomerate and therefore display post-Triassic movement. Downthrow on these mainly normal faults is consistently to the SW (Fig. 8). The amount of throw on these faults is variable but one which affects the Woodhill Bay Conglomerate in the middle of its outcrop reveals a displacement of 11 m (Pick 1964b). It is interesting to note at this locality the differences between the Triassic and Woodhill Bay Conglomerates where they are directly in contact. The characteristics of the latter are described at the next locality but the salient features of the Triassic Conglomerate are worth studying at this point.

The Triassic Dolomitic Conglomerate reaches its maximum thickness of 60 m at Ladye Bay but normally it is only 10 to 15 m thick. This deposit, more accurately described as a calcareous sharpstone conglomerate, can contain cobble and boulders up to 1.5 m long axis dimension. These large clasts are usually very angular and are dominated by Carboniferous Limestone and Old Red Sandstone debris. The low angle of dip, between 3° and 7° to the NW probably represents the original depositional dip of the sediment. The Conglomerate, therefore, comprises material which has undergone very little transportation. Gravitational forces, in addition to water transport, have played a major role in the sedimentation of this deposit which probably accumulated as alluvial fans, and/or scree deposits, at the base of a mountain front, and thus represents a 'typical' post orogenic sediment body deposited in an intermontane basin environment.

This last locality (D) at ST 4577 7679, near the steps and small stream, should be studied in order to examine the Woodhill Bay Conglomerate, its contact with the underlying Black Nore Sandstone, and the sedimentary features of the Sandstone. Here the Woodhill Bay Conglomerate is clearly exposed throughout its full thickness of 4 to 4.5 m. Generally the pebbles and cobbles in this polymictic roundstone conglomerate appear unsorted but there is a local tendency for the larger clasts to occur near the base of the formation. The clasts may be as large as 150 mm long axis dimension but the average size is of the order of 20 mm. The clasts are mainly composed of vein quartz and dark red quartzites but Wallis (1927a) also recorded chert, jasper, lithic sandstone, quartz schist, mica schist and spilite testifying to the extraformational origin of this conglomerate. The clasts are generally well rounded and measurement of some sixty pebbles shows the assemblage to be dominated by spherical and discoidal particle shapes. Sphericity values are high between 0.5 and 0.9. Imbrication of the pebbles within the Conglomerate is rare or absent but occasionally a preferred orientation of clasts parallel to bedding may be discerned. The pebbles and cobbles are held in a coarse sand matrix which may be bonded by either silica or carbonate cement.

Within the Conglomerate there are rare coarse grained sandstone lenses which are planar cross bedded, the cross bedding, according to Pick (1964a), indicating sediment dispersal towards the ESE. Wallis's (1927a) work on the petrography of the contained pebbles led him to believe that the source of the majority of the clasts was the Precambrian Mona Complex of Anglesey, and in particular the Gwna Group of the Complex.

The contact of the Woodhill Bay Conglomerate with the Black Nore Sandstone is best seen at this locality. No positive evidence of the Lower Old Red Sandstone age of the Black Nore Sandstone has yet been obtained but Kellaway & Welch (1955) suggest that a break in the sequence is indicated because of the presence of the Woodhill Bay Conglomerate at the base of the Portishead Beds. Pick (1964a) also states that evidence of a depositional break is present at the base of the Conglomerate but he regards the boundary as a disconformity.

In the field the base of the Woodhill Bay Conglomerate is seen to be irregular and uneven. Marked scouring of the underlying Black Nore Sandstone is obvious and infilling of irregular scour hollows by pebbles of the Conglomerate is apparent. However, there are very many erosion surfaces, of varying magnitude, present throughout the entire succession of the Old Red Sandstone in Portishead. The actual time gaps represented by these scoured surfaces are not known but it is probable that no large time intervals are involved.

The junction of the Conglomerate with the Black Nore Sandstone warrants further study and, in particular, the pedogenic limestone occurrences at this horizon should be investigated. Throughout the entire thickness of the Portishead Beds – with the exception of the Conglomerate at the base – no limestone horizons are encountered. However as soon as the Black Nore Sandstone lithologies crop out in the section these pedogenic limestone (calcrete) horizons abound. At the contact between the Conglomerate and the Sandstone much limestone is evident and, in addition to concretions, large pipe-like structures filled with limestone are seen within the Woodhill Bay Conglomerate. Upon close examination it can be seen that isolated pebbles of the Conglomerate are apparently contained in, and completely surrounded by, the limestone in the pipes. This calcretization is consistent with the pedogenic carbonates occurring in the Black Nore Sandstone. This could indicate that there is no great time interval across this contact and that the main unconformity is at the top of the Woodhill Bay Conglomerate and not at the base as has been previously suggested.

Below the Conglomerate to the SW of the steps good exposures of the Black Nore Sandstone are seen. Outcrops of this formation occur from this point some 3 km SW to Charlcombe Bay. In many places over this distance the

Black Nore Sandstone is totally obscured by the Triassic cover. The sandstone exhibits lithologies and sedimentary features consistent with an alluvial origin, the lack of siltstones in the sequence reflecting the dominance of channel fill sedimentation throughout its entire thickness. The most frequently occuring lithology in the Sandstone is fine to medium grained red quartzitic sandstone. Locally carbonate cement is present and horizons rich in muscovite can be seen. Practically all the sandstones have a lower erosional surface and internally the sandstones are cross bedded or flat bedded. The cross bedding indicates dispersal towards the SE (Pick 1964a). Intraformational conglomerates, which comprise mud flakes and rolled carbonate debris, usually overlie directly the scoured surfaces. Extraformational conglomerates, with vein quartz pebbles, and siltstones are subordinate in the sequence at this locality. Pedogenic limestone ('cornstones' of earlier authors) horizons are numerous at this point in the succession; in places the nodules appear dolomitized and at other points the carbonate has been weathered out to give the sandstone a cavernous appearance. All these salient depositional features of the Black Nore Sandstone can be studied at this locality and as the succession is traversed downwards towards Black Nore Point the repetition of the main lithotypes and sedimentary structures can be clearly seen.

The return journey can be made by climbing the steps and taking the path across the downs.

4
Volcanic rocks of the Bristol region

D. L. Speedyman

Objectives

In the Bristol region there are volcanic rocks of Silurian and Carboniferous ages, but they are very rare. This excursion describes some of the localities where they may be seen.

The outcrops of the Silurian rocks are some distance from the Carboniferous outcrops, and it is not intended that all the localities should necessarily be visited on the same excursion.

Special Features

Moons Hill Quarry is a working quarry with many unstable rock faces. Permission to visit the quarry must be obtained from John Wainwright & Co. Ltd., Downside, Shepton Mallet, Somerset.

Stoke Hill Basalt Quarry is also a working quarry and permission to visit it must be obtained from The Manager, Stoke Hill Basalt Quarry, Stoke St. Michael, near Shepton Mallet.

The localities at Middle Hope and Spring Cove are both coastal sections, and can only be properly studied at low tide. The coastal road between (ST 311 626 and ST 332 632) is a toll-road during the summer months.

Maps

Ordnance Survey 1:50 000 Sheets 182 (Weston-super-Mare & Bridgwater)
183 (Yeovil & Frome)
1:25 000 Sheets ST 36 and ST 64
Geological Survey One-Inch Sheets 263 (Cardiff) and 281 (Frome)

Main References

Boulton (1904); Morgan & Reynolds (1904); Reynolds (1907b, 1916a)

Outline Geology

Volcanic rocks in the form of lavas and tuffs occur sporadically in the Silurian and Carboniferous successions in the Bristol region. Each of these will be considered in turn.

Silurian volcanics Volcanic activity in Britain during the Silurian was virtually negligible, and one of the few places where Silurian volcanic rocks are preserved is in the eastern part of the Mendip Hills. The Mendips principally consist of a number of *en èchelon* periclines of Carboniferous Limestone, each with a core of Old Red Sandstone. The easternmost pericline also contains some Silurian volcanic rocks, which form an inlier surrounded unconformably by Old Red Sandstone rocks.

These volcanics consist of lavas of andesite and rhyodacite, tuffs, and agglomerates, their age having been determined by the presence of Silurian fossils in the bedded tuffs (Reynolds 1907b, 1908, 1912b). Their petrography and geochemistry have been described by Van de Kamp (1969).

They can be seen in two working quarries near Shepton Mallet, Moons Hill Quarry and Stoke Hill Basalt Quarry. Sunnyhill Quarry, which was described by Reynolds (1907b), is now overgrown.

Carboniferous volcanics Evidence of Carboniferous volcanic activity in the region has come from a number of outcrops to the south-west of Bristol.

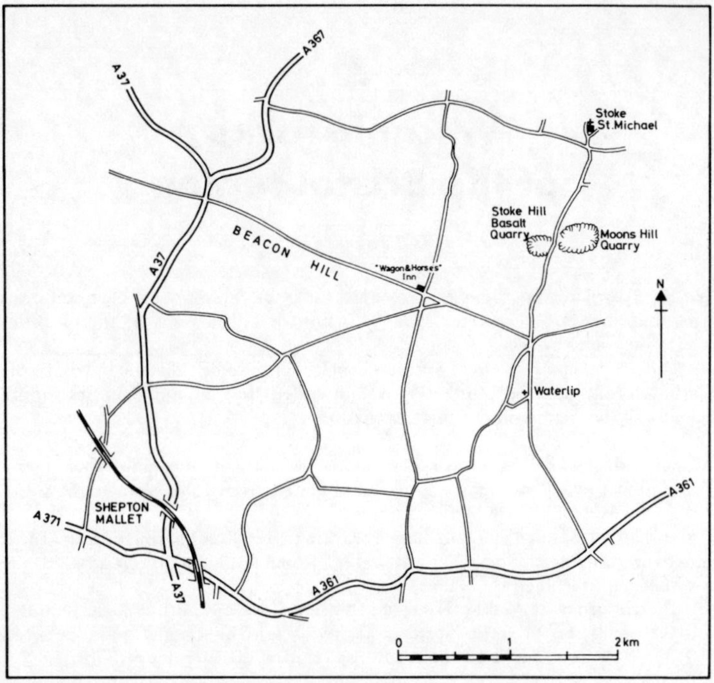

Fig. 1 Locality map for Silurian volcanics

These have been described by Reynolds (1904, 1916a) and a locality map is presented by Reynolds (1904). The inland exposures are now mostly overgrown, and the only localities worth visiting are the coastal ones at Middle Hope and Spring Cove.

On the Middle Hope promontory, Carboniferous Limestone of the Black Rock Group contains a sequence of volcanic rocks (Matthews *et al.* 1972), which has been described by Geikie & Strahan (1899) and Reynolds (1904, 1916a). The succession reveals a gradual onset of volcanic activity indicated by the presence of tuff horizons alternating with limestones, culminating in considerable depositions of tuff and the extrusion of a pillow lava, followed by a gradual waning of the volcanism as the limestone deposition progressively reasserted itself.

At Spring Cove, the Carboniferous Limestone is younger than the Black Rock Group, and contains a thick basaltic pillow lava with which is associated a coarse agglomerate. Descriptions of this locality have been given by Geikie & Strahan (1899) and Boulton (1904).

Itinerary 1: Silurian Volcanics

The two localities described are adjacent quarries at Stoke St.Michael, near Shepton Mallet (see Fig. 1).

1 **Moons Hill Quarry** (ST 662 461) This quarry lies on the east side of the road leading southwards from Stoke St. Michael, and vehicles can be parked at the main entrance.

The main rock-types in the quarry are greenish-grey andesite and rhyodacite. They are considerably jointed, sheared, and veined by calcite and green epidote. The bedding is indicated by a number of tuff horizons passing through the quarry, and is subvertical. One tuff horizon is well seen in the cliff on the left of the track descending the quarry, immediately below some of the quarry buildings. It forms a vertical band about 3 m wide in which soft, dark purplish tuff has been sheared, and veined by calcite parallel to the bedding. A second tuff horizon occurs just to the north. A detailed description of the quarry is not given because its morphology is constantly changing as the working progresses.

2 **Stoke Hill Basalt Quarry** (ST 661 460) This quarry is situated on the west side of the road leading southwards from Stoke St. Michael, virtually opposite Moons Hill Quarry.

In spite of the quarry's name, the igneous rocks here are the same as those in Moons Hill Quarry, i.e. andesites and rhyodacites. They can be seen in the lower levels of the quarry. Along the topmost level, around the western margin of the quarry, tuffs and agglomerates overlying the lavas can be seen. In places the agglomerate is coarse, containing blocks several centimetres across, some of which consist of earlier agglomeratic material. Again, a detailed description of the quarry is not given because it is a working quarry.

Itinerary 2: Carboniferous Volcanics

The principal exposures are at Middle Hope and Spring Cove, both in the vicinity of Weston-super-Mare (see Fig. 2).

1 **Middle Hope** Volcanic rocks occur at four exposures on the northern coast of Middle Hope, a hilly promontory at the north end of Sand Bay, north of Weston-super-Mare. The easternmost exposure, however, now lies within a Ministry of Defence mine-testing establishment, and will not be described.

Cars may be parked in the car park at the far end of the Sand Bay coast road, at the foot of Middle Hope (ST 3305 6596). Coaches must park in the car park by the bus stop half-way along the coastal road (ST 3320 6465).

To reach the **western exposure** go through the gateway in the lane leading to the café, just past the entrance to the car park, and immediately inside the gateway take the path on the left which goes steeply up the hillside. At the top, negotiate the stile and bear left across the grass towards a stone wall. Keep to the north of the wall and follow it as it runs parallel to the coast, right down to the shore below Swallow Cliff (ST 3245 6605).

At this locality the rocks have a strike of 070°, and a dip of 30° towards the south. The succession is described below, in ascending order, and is represented diagrammatically in Fig. 3.

(1) At the base of the succession, on the seaward side, are highly fossiliferous grey limestones containing many crinoids. The limestones have been churned up by burrowing organisms. Towards the top a reddish colour develops in places, possibly a precursor of the succeeding volcanism.

(2) 2 m of soft brownish fine-grained tuff, cut by numerous thin calcite veins.

(3) 40 cm of fine-grained well-laminated limestone, undisturbed by burrowing, but with slumping. Calcite veins having a suborthogonal pattern with a third direction bisecting the angle between them can be seen on bedding surfaces.

(4) 1 m of fine brownish tuff similar to (2), with calcite veins.

(5) 30 cm of well-laminated fine limestone like (3), with thin intercalations of brown tuff a few centimetres wide.

(6) 2 m of brown tuff with intercalated fine limestones a few centimetres wide. The upper part of the tuff contains dark lapilli of lava, occasional small blocks of shelly limestone, and broken shell debris, mainly crinoidal. Some of the calcite veins are flanked by dark metasomatic reaction zones.

(7) The tuffs grade rapidly into limestone about 1 m wide, which has been burrowed, and contains much shelly material including lamellibranchs, and fragments of tuff. There are thin intercalations of laterally-impersistent tuff

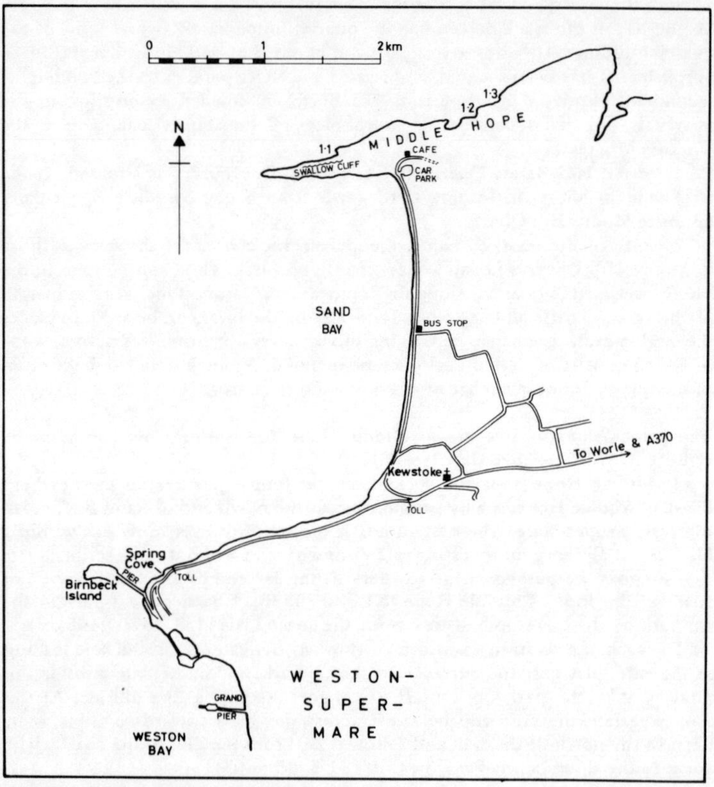

Fig. 2 Locality map for Carboniferous volcanics

containing shell debris.

(8) 3 m of soft brown tuff, a little coarser, and containing numerous dark lapilli and considerable amounts of shell fragments. There are numerous calcite veins predominantly subparallel to the bedding. Some of these veins form small thrust-nappe structures where they have deformed in a brittle manner, while the tuff has responded in a more plastic way.

(9) 5 m of darker brown soft tuff virtually devoid of calcite veins, but rich in lapilli of various lithologies and remains of shells, including colonies of corals overwhelmed by the tuff.

(10) 30 cm of sandstone.

(11) 70 cm of interbedded sandstones and purplish-brown tuffaceous siltstones showing cross-bedding and slumping.

(12) 4 m of massive highly-fossiliferous grey limestone rich in corals. Calcite is common in veins and cavities. At the extreme eastern end of the exposure, the limestone is offset about 2 m by a fault.

On the top-most bedding surface of this massive unit, a number of calcite-breccias can be seen. These penetrate down subperpendicular to the bedding, and one can be traced into the underlying siltstones and sandstones of units (11) and (10). They form irregular zones up to about 30 cm across, and consist of angular, disorientated fragments of finely-laminated limestone set in a matrix of calcite crystals. The limestone fragments have sharp margins,

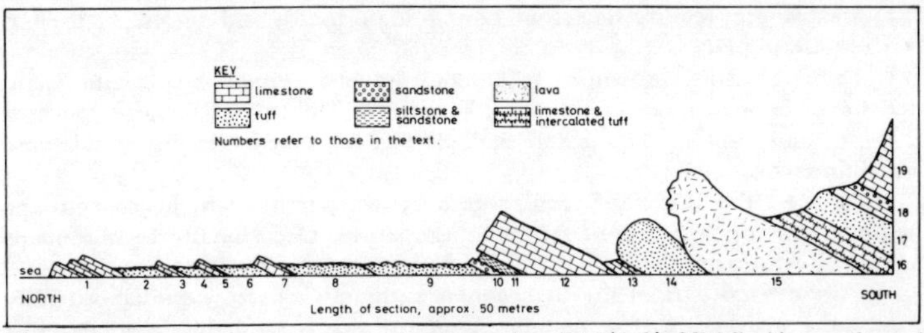

Fig. 3 Diagrammatic succession at western locality (1.1), Middle Hope, viewed from the west

do not touch each other, and in places the calcite matrix has prized open the bedding. The calcite of the matrix is generally fine-grained at the zone margins, and tends to be coarse among the fragments where volatiles would have been trapped. They probably represent fault breccias which have been intruded by calcite.

(13) 1 m of finely-laminated limestone at the top of which are purplish-brown tuffaceous siltstones.

(14) 5–10 m of greenish tuff, coarse and agglomeratic in places, with many dark lapilli. Calcite veining is very evident, and is predominantly subparallel to the bedding. In the upper part of the tuff there are a number of thin red cherty horizons.

(15) Massive basaltic pillow lava with a maximum thickness of about 5 m, net veined by calcite. The base of the lava, where it ploughs into the tuffs, can be seen well at the bottom of a high prominent rock standing up from the beach. On the face of this rock there are numerous pillows, especially near the base, and on either side of it there is a small fault which can be traced up into the cliff behind.

Towards the east, the cliff is essentially of tuff in which a few large pillowy bodies of lava, up to a few metres in length, can be seen. It is clear that the lava thins towards the east, suggesting that its source was in the west.

The upper part of the lava was more vesicular, for on the west side of the prominent rock, where the top of the flow can be seen, there are large calcite-filled amygdales up to 8 cm across, some of which are elongated parallel to the flow direction. The top of the lava can be traced over the large rock, where the fractured tops of the pillows are displayed, and the overlying sediments are seen penetrating down between pillows.

The lithology succeeding the lava is greenish tuff in some places, and cherty limestones in others.

(16) 3 m of bluish-grey cherty fossiliferous limestones with intercalated tuffs.

(17) Transition into 3 m of greenish tuffs containing abundant shell fragments.

(18) 2 m of reddish crinoid-bearing limestones and red ferruginous sandstones.

(19) Massive fossiliferous limestone of the cliff.

For an account of the capping Pleistocene deposits see Excursion **16,** loc. **11a.**

To reach the **middle exposure** follow the coast eastwards for about 1 km, past the first small pebbly bay, to a stone wall. Climb the wall by means of a wooden ladder, continue for a further 150 m, and descend to the next small westerly-facing bay (ST 3375 6645). The volcanics have been offset northwards by faulting, so they appear again in this bay.

The bedding here is steeper, being 45° with a strike of 085°. The succession is again described in ascending order, from the seaward side.

(1) Massive grey cherty limestone containing crinoids and corals. At the top it is reddish in places.
(2) 15 m of soft brownish tuff, considerably veined by calcite, with occasional beds of limestone about 20 cm thick. The tuff gets coarser towards the top, and contains lapilli and shell debris, especially crinoids, corals, and lamellibranchs.

In places, the tuff has irregular dark brown patches which cross-cut the bedding and appear to result from metasomatism. Occasionally fossil remains are partly within and partly outside the dark metasomatized patches.

In the central part of the tuff sequence, there is a narrow eroded-out gully containing crushed tuff and some hematite; it is probably a small fault parallel to the bedding.

Just over 1 m above this fault a narrow band of limestone shows slight folding.

Above this limestone band the tuff becomes coarser and contains many lapilli and shell fragments as well as diffuse lenticular calcareous patches. In places cross-bedding can be seen. Upwards, the tuff becomes increasingly coarser, and contains rounded pebbles, mainly of limestone and sandstone.
(3) 1.5 m of conglomeratic tuff with pebbles up to 25 cm across, which are held apart by the matrix. Lapilli and fossil debris still abound in the tuff.
(4) 80 cm of fine brown tuff at the base of which is a lenticular limestone.
(5) 80 cm of purple and grey finely-laminated sandstones and siltstones, at the base of which there is ripple-drift bedding.

In the central part there is an abundance of closely-spaced, narrow, vertical pipe-like tubes perpendicular to the bedding. They are worm burrowing-tubes and begin on numerous horizons, penetrating down for several centimetres through many layers. The bedding curves down towards each tube as a result of the feeding by the worms.

At the top of this unit, there are coarser burrowing features.
(6) 2 m of reddish-brown coarse tuff with lapilli.
(7) 4 m of massive grey coral limestone with a conglomeratic base. At the top are a few beds of siltstone, which are folded and have small-scale nappe structures.
(8) 3 m unexposed; this may represent weathered tuffs or a fault.
(9) Massive grey burrowed limestones, rich in well-preserved crinoids. It has been considerably replaced by chert in bands and lenses.

The pillow lava which occurred at the western exposure does not appear at this exposure, confirming that it thins and dies out towards the east. Furthermore, the thickness of the tuff sequence is also reduced towards the east, again suggesting that the source of the volcanics was to the west.

To reach the **eastern exposure** follow the coast further to the east for about 200 m, and descend to the next small bay (ST 3390 6655).

The volcanic sequence here is reduced compared to the middle exposure, and again the pillow lava is absent. The volcanics have been offset towards the north by a fault, which can be seen in the south-western corner of the bay.

This fault separates massive grey cherty crinoidal limestone to the west, from yellowish-brown tuff. The tuff is coarse and contains dark lapilli and fossil debris, especially crinoids, corals, and lamellibranchs. Calcite veins are very abundant. In the fault zone, thin limestone beds within the tuff have been stretched into deformed lenses, while the tuff has responded in a more plastic way. There is also some metasomatism of the tuff in the fault zone.

About 15 m east of the fault, a second small fault within the tuff can be seen continuing up into the cliff towards the first fault. Eastwards from this little fault the tuff is pebbly and very fossiliferous, and is cut by many calcite veins. About 10 m from the little fault the tuff contains a bed of nodular limestone about 40 cm thick. This limestone has coarse jointing perpendicular to the bedding, with slight faulting at some of the joints.

The tuffs overlying this limestone appear to lie in the core of an open syncline plunging towards the SE. Just to the east, the nodular limestone occurs again, on the other limb of the fold, dipping in a different direction. This massive limestone has deformed in a brittle manner, fracturing at the fold hinge, whereas the tuffs in the core have responded in a more plastic way. There is considerable calcite veining in the tuffs parallel to the axial plane of the fold. At the top of the rocks, disharmonic folding in the core of the fold is demonstrated by the different attitutes of limestone beds compared with the massive unit already described.

On the NE limb of the syncline, just above the massive nodular limestone, there are zones of *en èchelon* sub-sigmoidal tension gashes in the tuff, which have remained unfilled. Towards the NE, the 40 cm nodular limestone is apparently cut off and repeated by faulting. The tuffs here contain numerous calcareous nodules and diffuse lenses rich in shelly material. Continuing down the succession toward the north, the tuff contains thin beds of nodular limestone. A massive unit of burrowed limestone 1 m thick, containing crinoids and corals, is underlain by 40 cm of purplish siltstones. Then 8 m of yellowish-brown fine tuff with thin nodular limestones is underlain by massive grey cherty limestones containing crinoids and corals. These limestones continue to the sea.

2 **Spring Cove** Spring Cove is situated on the coast 250 m NE of the pier leading to Birnbeck Island, which lies to the NW of Weston-super-Mare.

Vehicles may be parked on the roadside either 100 m west of the toll gate (ST 3100 6255) or 100 m east of the toll gate (ST 3120 6265). Alternatively, they may be parked at the entrance to the pier (ST 3090 6240). Go through the railings beside the footpath at a point 50 m west of the restaurant by the toll gate. Descend the short path to the cliff-edge; turn left along a path leading down into the eastern end of Spring Cove (ST 3100 6255).

The lower part of the path follows a fault parallel to the bedding, in which the rocks have been crushed and there are traces of hematite. On the seaward side of the fault, the rocks are shelly limestones containing occasional clastic or neptunian dykes of reddish-brown siltstone, penetrating down into the limestone. To the south of the fault, the rocks are reddish-brown sandstones and siltstones containing large corals up to 6 cm across. The bedding dips at 30° with a strike of 105°.

The red siltstone group is about 3 m thick, and grades up into 9 m of reddish-brown highly fossiliferous oolitic limestone. It is full of small shelly fragments, and some beds have concentrations of fossil debris, especially crinoids. The ooliths have a purplish-brown coating, possibly volcanic mud, and the matrix may contain tuffaceous material. Faint cross-bedding can be seen in places in the cliff. There are numerous veins of calcite. This oolite is overlain by 4 m of hard limestone, grey in the lower part, and tending to become red and more ferruginous towards the top. Above this limestone is a purplish-brown basaltic pillow lava, with an irregular base, and net-veined by calcite. Pillows with cracked crusty rims are best developed at the base, and are up to a metre or more across. As the base of the lava is followed westwards, large concavo-convex pillows can be seen to sag down between underlying pillows, where they were piled up as the lava was extruded on the sea-floor.

Continue to follow the base of the lava westwards, until the exposures of the underlying limestone end in the beach boulders. In the cliff above there is a zone about 5 m wide of agglomeratic tuff, cutting obliquely across the lava and reaching down to the beach. This zone contains large rounded pillow-like masses, some several metres long, of lava set in an agglomeratic matrix. There are also blocks of limestone in the agglomerate, and some of the basalt pillowy masses contain irregular inclusions of limestone. The agglomeratic matrix consists mainly of densely-packed angular fragments of the lava, and is probably an autobreccia. The blocks of limestone and pillowy lava are often closely spaced, but rarely touch one another. They tend to be elongated parallel to the margins of the zone. In places the matrix has a slight planar fabric which is generally parallel to the zone itself, but swings around the lenticular included blocks. This fabric is accentuated by calcite-veining along this direction. This agglomeratic autobreccia is probably the result of a submarine slide, in which pillowy lava containing limestone blocks slid off the top of the flow, became autobrecciated, and came to rest on the sea floor.

Above this 5 m zone, the succeeding lava is still somewhat agglomeratic, but the proportion of autobreccia matrix is smaller, and the planar fabric is less marked. There are larger volumes of coherent pillow lava, with occasional narrow irregular agglomeratic zones. The top of the lava flow can be seen in a gully at the foot of the cliff. Here the lava is strongly pillowed, and is amygdaloidal with calcite-filled amygdales up to 1 cm across. The total thickness of the flow, including the agglomeratic zone, is about 15 m. Overlying the lava is about 1 m of tuffaceous limestone, cross-bedded in places, above which the cliff consists of massive grey shelly limestone, with stylolites.

Towards the west, the whole thickness of the flow becomes agglomeratic. Blocks of limestone and pillows of lava, some of which are amygdaloidal, are held apart by the agglomeratic autobreccia matrix. In the cliffs here, some long narrow clastic dykes can be seen, in which reddish-brown siltstone penetrates down fissures in the linestone. Further west, the flow becomes a massive coherent purplish-brown basalt lava containing olivine pseudomorphs. It is somewhat pillowy and amygdaloidal in places, particularly in the upper part. This continues westwards for about 100 m until it disappears under beach boulders. The length of the whole section is approximately 200 m.

5
The Carboniferous Limestone of the Mendip Hills

R. J. G. Savage

Objectives

The purpose of the excursion is to examine the Lower Carboniferous succession on the Mendip folds, noting the facies variations and palaeontological succession. There are three itineraries, to Burrington, Cheddar and Ebbor.

Special Features

Care must be exercised on the combe and gorge faces to avoid injury from rock falls. The sites are in popular tourist areas and coach parking places are limited. During the summer months coaches may only descend Cheddar Gorge and not ascend.

Maps

Ordnance Survey	One-Inch Sheet 165 (Weston-super-Mare)
	1:50 000 Sheet 182 (Weston-super-Mare & Bridgwater)
	1:25 000 ST 45, ST 54
Geological Survey	One-Inch Bristol District Sheet

Main References

Green & Welch (1965); IGS Memoir 280 (Wells & Cheddar).

Outline Geology

In early Carboniferous times the seas transgressed north-eastwards across the Bristol area over the Old Red Sandstone shore lines. For the most part carbonate facies were deposited, ending in the sands of the Millstone Grit times. Other excursions (6 & 7) examine the higher Coal Measure facies and flora. The karst features of the Carboniferous Limestone are examined in excursion 5 and the igneous facies in excursion 4.2. A detailed account of the geology will be found in the Wells and Cheddar memoir.

The Carboniferous Limestone is well represented around the Bristol district, but nowadays opportunities to examine it are limited. The Avon Gorge, so long the classic locality, is no longer safely accessible to parties. Many of the small Mendip quarries have closed down and become overgrown; a few quarries have become enormously large with outputs in excess of 8,000 tons per day. Neither type is ideal for study purposes and on this excursion natural exposures in combes and gorges will be visited.

Limestones dominate the Lower Carboniferous succession, with dolomites, calcite mudstones and shales. The calcareous facies vary from chinastones through oolites to coarsely crinoidal limestones. For two reasons the Folk nomenclature is not used; the author has not personally sampled all the rock to determine their type, and the classification, while ideal in the laboratory, is not often easily applicable in the field.

Around the first decade of the century Vaughan (1905) zoned the Carboniferous Limestone on the basis of their faunas, taking the Avon Gorge as his type section, the stage being named the Avonian (Reynolds & Vaughan 1911). They recognised five zones and a series of subzones, each characterized by a brachiopod or coral. The zones were given initial letters corresponding to the zonal coral or brachiopod, except for the first whose name has now been

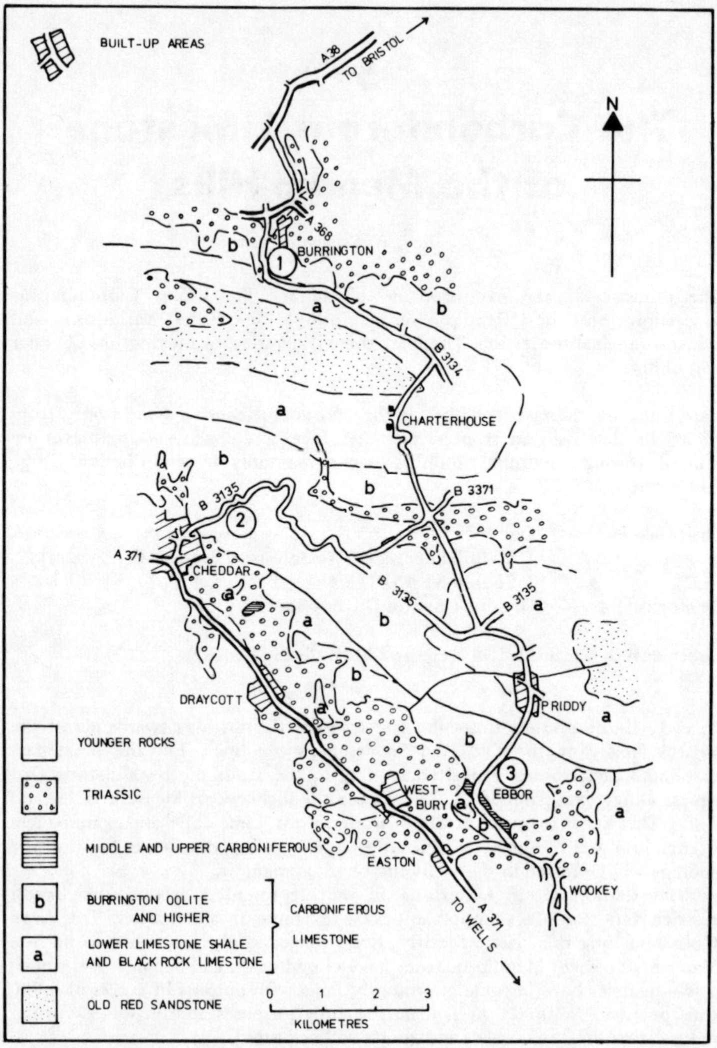

Fig. 1 Geology map of the Mendip hills with three itinerary areas

changed.

Zone	Zonal Fossil
D	*Dibunophyllum*
S	*Seminula*
C	*Caninia*
Z	*Zaphrentis*
K	*Cleistopora* (now *Vaughania*)

The zonal scheme can be applied to other sections in the Bristol district in Somerset and Gloucester, and across the Bristol Channel in South Wales. However there are difficulties. The zonal fossils were not exclusively restricted

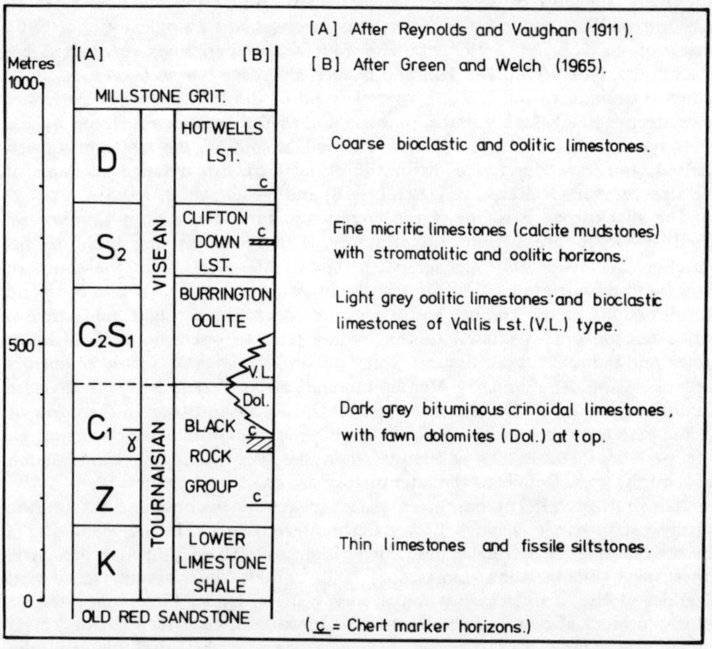

Fig. 2 Lower Carboniferous stratigraphic succession and nomenclature

to their zones; they might appear in an earlier zone or survive into a later zone. Nevertheless, with experience, they, and in particular the subzonal species, provide a very useful and on the whole reliable guide to stratigraphic horizons.

In the course of resurveying the district, officers of the Institute of Geological Sciences devised a different scheme based on facies; their rock units are shown in Fig. 2. Again the nomenclature derives mainly from the Avon Gorge, e.g. Black Rock Limestone, Clifton Down Limestone and Hotwells Limestone. The facies can mostly be traced through the Mendips, though the names change – Cheddar Limestone, Burrington Oolite etc. When the two schemes are compared it is seen that their units are similar but not identical – a reflection of the fact that the 'zonal fossils' are in part facies fossils. It should be remembered that the facies nomenclature identifies lithostratigraphic units and the boundaries are inherently diachronous. The fossil nomenclature identifies biostratigraphic units using organisms that are at least in part facies dependent and suffers the same limitations. Neither scheme is completely satisfactory and for this reason both are used in this account. The alternative is a chronostratigraphic division based on planktonic faunas. Goniatites are used in deeper water facies of the Carboniferous but are so rare in the shallow water epicontinental Carboniferous as to be of little value. Conodonts hold out hopes of fulfilling the need but the detailed results are not yet available. Even when they are, microfossils require laboratory study and are of no use in field identification.

During Late Carboniferous and Permian times the region was affected by Variscan orogenic movements. The northern limit of the front or mountain belt coincides essentially with the line of the Mendip hills, with to the south more intense folding, cleavage and overthrusting (e.g. in north Devon) and in the foreland to the north less marked effects. The Variscan front was essentially aligned east-west, but was deflected on meeting the Lower Severn axis and the Bath axis; the resultant structure was not a single anticline but a series of *en échelon* periclines. The four major periclines recognised are Blackdown, North Hill, Pen Hill and Beacon Hill; each has an E–W axial trace which is oblique to the NW–SE overall trend of the fold belt. The periclines have steeper north facing limbs, perhaps due to the resistance offered by the N–S synclinal Coal Measure basins. As well as folding, the strata have been faulted, and have developed distinct joint patterns. For detailed accounts of the structures see Kellaway & Welch (1948) and Welch (1929, 1933).

The Blackdown pericline is the largest and both Burrington Combe and Cheddar Gorge are within this structure. The north dipping limb of this pericline is considerably steeper (with dips of 60°–70°) than the southern limb (with dips around 20°). Joints in the limestones strike E–W and N–S and the directions of the combes and gorges are controlled by these joint trends. Major tension cracks related to the folding tend to open along E–W hinge zones and many of these fissures were later infilled with Mesozoic sediments (see excursion 9). Similarly Mendip mineralization occurs in veins that are predominantly E–W. Erosion has exposed Old Red Sandstone in the cores of all the periclines and on Beacon Hill even Silurian is exposed (see excursion 4). The periclines stood out as islands when the early Mesozoic transgression reached the area. Details of the later history are given in excursion 9.

Ramsbottom (1970) has given palaeogeographic reconstructions for the south-west province through Lower Carboniferous times. In the Mendip area the broad sequence is from shales (Lower Limestone Shales) through bioclastic limestones (Black Rock Limestone) and oolites (Burrington Oolite and Cheddar Oolite) to calcareous mudstones (Clifton Down Limestone), then a return to bioclastic limestones (Hotwells Limestone), each facies with a clear faunal assemblage. This may be interpreted as a gradual shallowing of the waters of the epicontinental sea with a final brief return to clearer water conditions before the appearance of the arenaceous facies of the Quartzitic Sandstones.

Itinerary 1: Burrington Combe

Burrington Combe is one of the most spectacular scenic features on the Mendip Hills, and exposes the finest sections in the lower Carboniferous, with almost a complete sequence through the Avonian (Tournaisian and Viséan). The beds strike approximately eastwest and dip north at angles around 60°; the northern limb of the Blackdown pericline is more steeply dipping than the southern. In the lower N–S part of the combe the dip section can be followed from the Hotwells Limestone to the Black Rock Limestone, younging southwards up the combe. The upper E–W part of the combe provides a strike section with the Great Scarp in the Black Rock Limestone. The lowest beds can be seen in the Twin Streams to the south of the combe. The thickness of the Lower Carboniferous in the combe is as follows:

	metres
Hotwells Limestone	30+
Clifton Down Limestone	160
Burrington Oolite	200
Dolomite	20
Black Rock Limestone	270
Lower Limestone Shales	150
	830

Joint patterns control the distribution of the two major parts of the combe and also minor features such as stream courses, cave geometry and scarp outlines. Dolomitic Conglomerate is deposited in the lower section of the combe and beyond the combe is a spread of Pleistocene – not normally

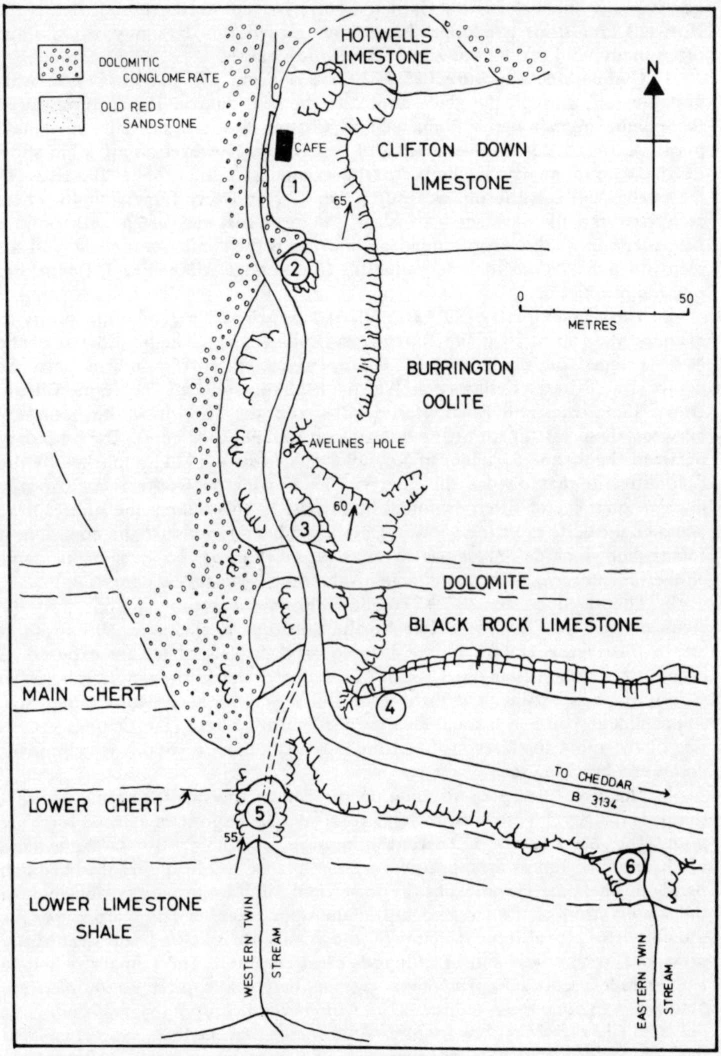

Fig. 3 Locality map for Burrington Combe

exposed. Excursion 6 should be consulted for the karst phenomena in the area.

The itinerary starts at the entrance to Burrington Combe and proceeds southwards up the combe; due to the northerly dip of the succession, the youngest beds are met with first and as the itinerary is followed, progressively lower (and older) strata are encountered. Dolomitic Conglomerate can be seen near the mouth of the combe opposite the Burrington Garden Centre (ST 4762 5895). The unconformable junction with the Carboniferous Limestone is not visible.

The highest Carboniferous beds in the combe are met with in the first quarry, now the Burrington Garden Centre. At the northern end of this the Hotwells Limestone is exposed in the rock garden; it is dark grey, massive and often shelly with productids and *Lithostrotion* masses.

1 **Combe side exposure** (ST 4770 5885) From the Garden Centre walk past the café through the glade almost to the next quarry. The eastern wall of the combe hereabouts is composed of Clifton Down Limestone, nowadays poorly exposed due to overgrowth of vegetation. However about 30 m short of the second quarry a climb up the combe side for about 20 m reveals reasonable and fossiliferous exposures. The Clifton Down Limestone (S_2) here comprises mainly dark grey to black calcite mudstones often with oolites. *Lithostrotion* masses are abundant; *L. martini* is the commonest coral, and the plentiful brachiopods include chonetids (*Chonetes papilionaceus*), *Composita* sp., and productids.

2 **The second quarry** (ST 4765 5875) This provides a good opportunity to examine the top beds of the Burrington Oolite ($C_2 S_1$). The boundaries of the two systems of nomenclature do not exactly correspond, as can be demonstrated here. Kellaway & Welch (1955) introduced the terms Clifton Down Limestone and Burrington Oolite, and the IGS draw the boundary between these just north of the quarry (not well exposed here). The boundary between the S_1 and S_2 zones of Reynolds & Vaughan (1911) is marked by the dolomitic rib that divides the quarry. The Burrington Oolite is a light grey massive oolitic and often crinoidal limestone. In this quarry the oolites have occasional calcite mudstone bands. Corals and brachiopods are the commonest fossils and include: *Palaeosmilia murchisoni, Lithostrotion martini,* large chonetids (*Megachonetes*), and large productids (*Gigantoproductus*).

3 **The third quarry** (ST 4770 5860) From the second quarry, past the Rock of Ages and Aveline's Hole, to the third quarry inclusive, all exposures are in Burrington Oolite. In the third quarry the basal beds are exposed. A dolomitized zone (20 m) just south of the quarry marks the junction between C_1 and C_2, and between the Burrington Oolite and the Black Rock Limestone. A prominent band in a small clearing just south of here (the *Caninia* bed) is one of the most fossiliferous horizons in the whole succession, with abundant corals and brachiopods.

4 **The Great Scarp** Continuing up the combe towards the corner where it sharply changes direction from N–S to E–W, the exposures at road level are poor. The Black Rock Limestone comprises predominantly dark grey and black crinoidal limestones, massive in some places and in others thin bedded. Bands of chert are common in the lower beds and are in part responsible for the preservation of the Great Scarp in the upper combe. The scarp slopes are usually richly fossiliferous; many of the fossils are silicified and beautifully preserved, with corals and brachiopods most common. The faunal list is long and includes: corals – *Amplexus* sp., *Caninia cornucopiae, Caninophyllum patulum, Cyathaxonia cornu, Fasciculophyllum* spp., *Syringopora* sp.; polyzoans (*Fenestella*); brachiopods – *Chonetes* spp., *Leptaena* sp.; *Productus* sp., *Schellwienella* sp., *Schizophoria* sp., *Spirifer* sp.; gastropods (*Bellerophon*) and pelycopods (*Acanthopecten*).

5 **Western Twin Stream** From the Great Scarp return to the corner and take the track south up the Western Twin Stream. The exposures along the track past Sidcot Swallet as far as Goatchurch Cavern are in the lower beds of the Burrington Oolite (Z beds). Chert bands cut through the limestone which is very fossiliferous, with crinoids, corals (often silicified), and brachiopods (*Spirifer* and *Chonetes*).

The shales of the Lower Limestone Shales (K beds) can be seen in the stream near the adit. The sequence comprises alternating shales and limestones, with a rich fauna of brachiopods (chonetids) and the zonal fossil *Cleistopora* has been found. Exposures are poor further upstream and it is best to walk up the Eastern Twin Stream to see the junction with the O.R.S.

6 **Eastern Twin Stream** Return to the corner of the combe and walk up the combe for about 300 m to a break in the scarp where the Eastern Twin Stream emerges. The stream disappears into the Black Rock Limestone at East Twin Swallet without entering the main combe. Going up the stream the Lower Limestone Shales are seen just before the first streamlet joins from the east side. Continuing up the main stream, the bridlepath over the downs crosses the stream near the base of the Lower Limestone Shale and soon after is seen a red band of highly crinoidal limestone, usually equated with the Bryozoa bed of the Avon gorge. Stratigraphically below this band is a very coarse crinoidal limestone, which is followed further upstream by alternating shales and crinoidal limestones. At 200 m O.D. can be seen the junction of the Carboniferous Lower Limestone Shales with the Old Red Sandstone; after rains when the stream is running fast this junction is difficult to locate.

Continue to the top of the combe and follow the signposted B 3134, B 3371 and B 3135 to the head of Cheddar Gorge. (Near the crossroads at the Miners Arms are the remains of former lead mines.)

Itinerary 2: Cheddar Gorge

Cheddar Gorge is certainly the most spectacular geological feature on Mendip, if not in the whole Bristol district. In consequence it suffers chronically as a 'beauty spot'; the area is riddled with tourists, coaches and motorcars. The walls disgorge boulders periodically making ascent perilous.

Cheddar is on the southern limb of the Blackdown pericline and the angles of dip (around 20° at 120°) are much shallower than on the northern limb (e.g. in Burrington Combe they are around 60°). So although the dip section in the gorge is longer than in Burrington Combe, fewer beds are exposed. As with Burrington, the beds young down the gorge. The itinerary starts at the top of the gorge and proceeds towards Cheddar village. The karst phenomena are described in excursion 6. The sequence of the Carboniferous Limestone in Cheddar is as follows:

		metres
Hotwells Limestone		37+
Clifton Down Limestone	106	
with Cheddar Oolite	53	194
and Cheddar Limestone	35	
Burrington Oolite		33+
		264

Five stops are itemised, one to illustrate each stratigraphic level. The sequence is essentially similar to that of Burrington Combe, save that the Clifton Down Limestone succession is thicker and its threefold subdivision is given a separate nomenclature by the IGS.

1 **Black Rock Gate** (ST 4820 5450) The Burrington Oolite (C_2S_1) outcrops for about 300 m from a point 50 m north of the gate to ST 4800 5460. It comprises crinoidal limestones and oolites, and the fauna includes *Lithostrotion* cf. *martini*, *Palaeosmilia murchisoni* and *Composita* cf. *ficoidea*.

2 **Reservoir Enclosure** (ST 4750 5450) The Cheddar Limestone (S_2) outcrops continuously between the last stop and the reservoir. The facies comprise black splintery limestones, calcite mudstones, crinoidal limestones and oolites. Many of the top beds are rich in fossils, including *Carcinophyllum*, *Lithostrotion*, *Palaeosmilia*, *Composita*, and *Productus*.

3 **Horseshoe Bend** (ST 4720 5435) From the reservoir to the far side of the Horseshoe bend there outcrops the Cheddar Oolite (S_2); this comprises grey-black coarse oolites, current-bedded and porcellaneous, with calcite mudstones and splintery limestones. The fauna includes *Composita*,

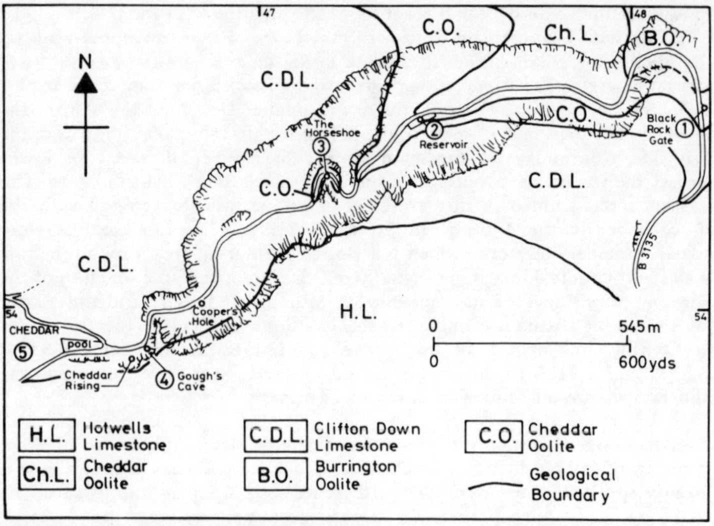

Fig. 4 Locality map for Cheddar Gorge

Lithostrotion, *Chonetes* and productids.

4 Gough's Cave (ST 4671 5390) The gorge section between the Horseshoe bend and Gough's Cave exposes the upper part of the Clifton Down Limestone (S_2). The facies are grey to black banded calcite mudstones with fine grained limestone and chert stringers. Algal beds occur and *Lithostrotion* is common.

5 Cheddar Village (ST 465 539) From Gough's Cave south through the village, the Hotwells Limestone (D) occurs high up on the southern cliff, but exposures are difficult of access. The grey crinoidal limestones contain *Lithostrotion* and productids. (See excursion **6**, loc. **14** and **15** for geomorphological features.)

Itinerary 3: Ebbor Rocks

From Cheddar village take the A 371 towards Wells along the southern fringe of Mendips with the splendid view southward across the Trias plain. The Mendip hills rise steeply to the north, fringed by a blanket of Dolomitic Conglomerate. At Draycott the conglomerate was formerly quarried as Draycott stone and is still to be seen in the buildings of Draycott and in Temple Meads railway station, Bristol. The view to the south across the expanse of the Trias plain is interrupted by occasional outliers of the Carboniferous: Nyland Hill opposite Draycott and Lodge Hill opposite Westbury-sub-Mendip. Beyond, a low NW–SE ridge of Rhaetic capped hills extends from Weare to Theale, while further away the parallel ridge of the Polden Hills mark the line of the Lower Lias outcrops, with two prominent outliers of Middle and Upper Lias: Brent Knoll to the west and Glastonbury Tor to the southeast (identified by the summit tower).

Cars may turn left in Easton along the minor road toward Wookey Hole; after 1.2 km at Ebbor Farm fork left up the steep hill and park in the Ebbor Reserve car park. Coaches must keep to the A 371 as far as Haybridge, 1 km short of Wells, and take the signposted road to Wookey Hole. From the coach park a walk of about 1 km takes the visitor to the Ebbor Nature Reserve. This is administered by the Somerset Trust for Nature Conservation, on lease from the National Trust to the Nature Conservancy Council. Hammers may not be used within the reserve.

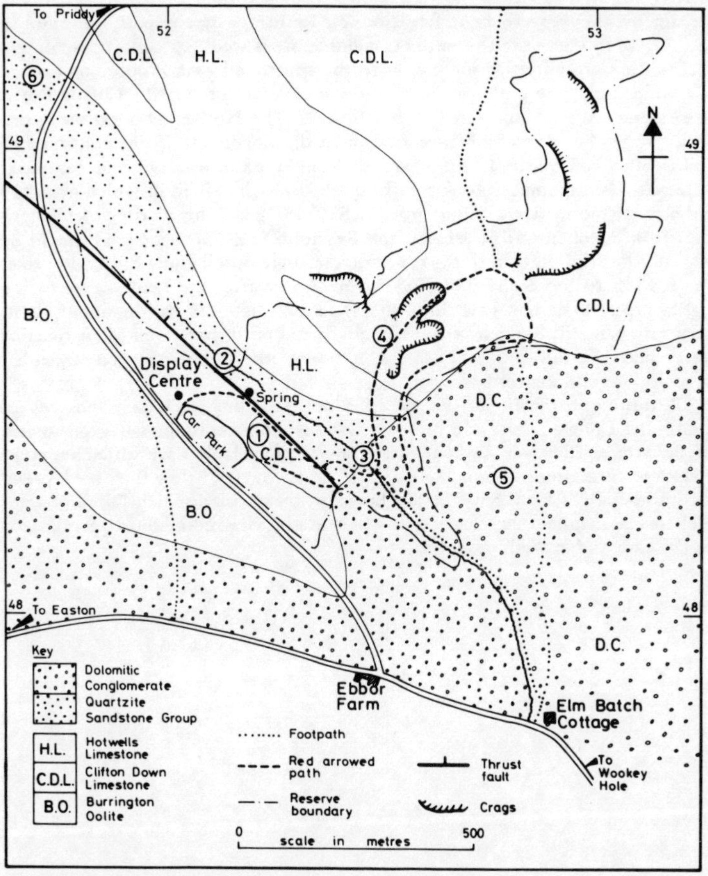

Fig. 5 Locality map for Ebbor Rocks

The Ebbor Rocks lie just west of the Pen Hill pericline. Here we see a sharp syncline with, to the south, a complementary anticline. The succession is in Burrington Oolite, Clifton Down Limestone and Hotwells Limestone, all with fairly low dips (between 20° and 40°) and NW–SE strike. They are succeeded by the Quartzitic Sandstone Group with shales and quartzitic sandstones; these have steeper dips around 70° as they disappear under the older Burrington Oolite and Clifton Down Limestone of the Ebbor Thrust. The thrust is part of the Cheddar-Wells thrust belt which thrusts northeastwards with a maximum upthrow near Cheddar of some 1000 m, decreasing towards Wells.

Enter the Reserve by the Display Centre and follow the red signposted route. About 100 m along the path a crag can be seen high up on the right (Fig. 5, loc. 1); this is the Clifton Down Limestone of the Ebbor Thrust over-riding the older rock below and is seen here to be highly fractured with seams of shattered chert. Continue along the path a short distance until a concrete lined pond is seen off to the left. This formerly filled from a spring at the junction of the limestone with the shales below. Only the spring is still visible but a little lower down the shales are exposed in the banks of the stream by the Reserve boundary (loc. 2). Returning to the path continue to loc. 3. Around here the Dolomitic Conglomerate is well exposed; it completely masks the Carboniferous succession to the south and east. Proceeding toward the gorge along the path good exposures are encountered in the Clifton Down Limestone with its fine scree slopes (loc. 4). The Hotwells Limestone is not easily accessible, forming higher ground to the northwest. If one continues to follow the red route it leads via the upper path back to the car park. Alternatively one may follow the footpath through a field in which over 200 pebbles of picrite were found (loc. 5, ST 528 483). The picrite is similar to that from Menheniot in Cornwall and Reynolds (1912a) suggested it might be glacially derived. Return to the car park, continue uphill along the Priddy road for 0.8 km to loc. 6 (ST 517 491). Immediately after the road swings to the right, cross a six-bar gate into the field on the left; within about 30 m quartzite can still be seen in the old quarries. Poorly preserved plant remains have been found and the shales between the quartzites have yielded brachiopods and gastropods.

Returning through the village of Wookey Hole, roadside exposures in Dolomitic Conglomerate can be seen; walls and houses in the village make use of the stone. The river Axe emerges from Wookey Hole Cave which has good stalactite formations and in the east wall of the ravine below the cave Hyaena Den and Badger Hole have in the last century yielded a rich fauna of Late Pleistocene mammals, a selection of which are on view in the museum adjacent to the cave and in Wells museum.

6

Limestone features and the geomorphological evolution of the Mendip Hills

D. Ingle Smith

Objectives

The aim of the excursion is to demonstrate the landforms that are developed on the Carboniferous limestones of the Mendip Hills. Additional sites have been selected that enable aspects of the evolution of the landscape and the hydrology to be illustrated in the field. The excursion falls into two parts. First, a detailed study on foot of the Burrington Combe area which is suitable for a half-day visit and, secondly, a traverse across central Mendip to Cheddar which can only be undertaken by the use of private transport. The complete excursion is suitable for a full field day.

Special Features

The itinerary includes a number of cave entrances but *under no circumstances* should any attempt be made to follow the cave passages underground. Should, on a separate occasion, a trip underground be contemplated it is essential that the party is small in number, adequately equipped and that experienced leaders accompany the party. There are a number of caving clubs on Mendip and a list of addresses is given in Barrington & Stanton (1972); most of the clubs will give advice and possibly provide qualified leaders for underground visits provided that adequate notice is given.

Many of the sites described in the itinerary are on common land or adjacent to footpaths and there is no need to seek permission to walk across the area provided all aspects of the Country Code are observed. Some of the other sites are on private land and while every attempt has been made to give details of where permission may be obtained, there is clearly no guarantee that such permission will be forthcoming.

Localities 4 to 8 can only be visited on foot. Cars or other vehicles can be parked without prior arrangement in old quarries some 50 m or so to the south of the café in Burrington Combe.

Permission to visit Wurt Pit (loc. 10) should be obtained from Mr. W. J. Masters, Ebbor Farm, Wookey Hole. Wells 73067.

Maps

Ordnance Survey	One-Inch Sheet 165 (Weston-super-Mare)
	1:50 000 Sheets 172 (Bristol & Bath)
	182 (Weston-super-Mare & Bridgwater)
	1:25 000 Sheets ST 45, ST55
Geological Survey	One-Inch Bristol District
	1:25 000 Sheet ST 45
Soil Survey	One-Inch Sheet 280 (Wells)

Main Reference

Smith & Drew (1975)

Outline Geology

An outline of Carboniferous Limestone geology is given under Excursion 5, pp. 55–64.

Itinerary 1 **Burrington Combe Mouth** (ST 482 600) The form of the periglacial fan associated with the mouth of Burrington Combe can be well seen from the road. The fan is composed of gravelly head material but is not normally exposed in the field. The nature of the superficial material results in a freely draining soil which is termed the Langford Series and this contrasts to the neighbouring soils, the Worcester Series, which are developed on the Keuper Marls. The details are discussed in Findlay (1965). The lighter Langford Series results in earlier grazing for livestock. The fresh morphological form of the fan would suggest a relatively late date for its formation and it would appear that the gravels originated at a phase of periglacial conditions when subterranean drainage ceased due to permafrost formation.

2 **Langford** (ST 4662 5930) Before continuing with the excursion into Burrington Combe it is convenient to visit the large spring at Langford. This is best approached along a footpath to the west of the stream from the hamlet of Upper Langford. Parking for motor coaches is difficult. The spring issues from a thin cover of gravels which overlie the Keuper Marls although the main mass of the limestone can only lie 1 to 2 m below these deposits. The mean flow is approximately 2 800 m^3/day. The swallets at Read's Cavern and Eastern Twin Stream are known to connect with Langford Rising; for discussion see Drew *et al.* (1968).

3 **Rickford Rising** (ST 4878 5916) An alternative spring site to loc. 2 is Rickford Rising which lies a few metres to the north of the A 368. Rickford Rising issues from the Dolomitic Conglomerate which has formed a thin cover on the Carboniferous Limestone. Even at times of low flow it has not proved possible for cavers to explore the spring for more than a few metres. The mean flow is some 5 500 m^3/day. The swallets at Read's Cavern, Western and Eastern Twin Stream and Ellick's Farm have been traced as feeders of the Rickford Rising. Further details of the hydrology and water chemistry are given in Drew *et al.* (1968) and Newson (1972). Some 500 m to the east of Rickford Rising and 10 m to the south of the main road is an intermittent spring known as Squire's Well (ST 4928 5908). This would appear to act as an overflow spring for Rickford Rising in times of high discharge conditions, see Tratman (1963). Langford and Rickford Risings form the major feeders for the reservoir at Blagdon Lake.

4 **Garden Centre** (ST 476 590) Exposures of Dolomitic Conglomerate that represent the infilling of a sub-Triassic forerunner of the lower part of Burrington Combe can be seen in the road cut opposite the Garden Centre and again in roadside sections some 50 m further up the Combe. The Dolomitic Conglomerate at these exposures is angular to sub-angular and includes fragments, mainly of Carboniferous Limestone, larger than 20 cm. Calcite crystals of a later date can be seen infilling voids in the conglomerate. The general nature of the sub-Triassic topography can be illustrated by considering the Triassic outcrops on the periphery and the plateau surface of the Mendips shown on the One-Inch or 1:25 000 geology maps.

5 **Aveline's Hole** (ST 4761 5867) Some 150 m further up the Combe on the western side of the road is the 'Rock of Ages' which is marked by a plaque. This cleft acts as a specific illustration of the differential weathering of individual beds within the main mass of the limestone, the cleft is clearly developed on a single bedding unit. For the energetic a scramble to the top of the Rock of Ages affords a good viewpoint to see the overall morphology of the lower part of the Combe.

Aveline's Hole is located a further 50 m along the road, on the eastern side of the Combe at road level. The cave was discovered in 1797 and the boulders blocking the entrance removed. Descriptions of the excavation make it clear that a large number of skeletons were found. Early excavation techniques and the disposal of the finds has meant that much of the material and information

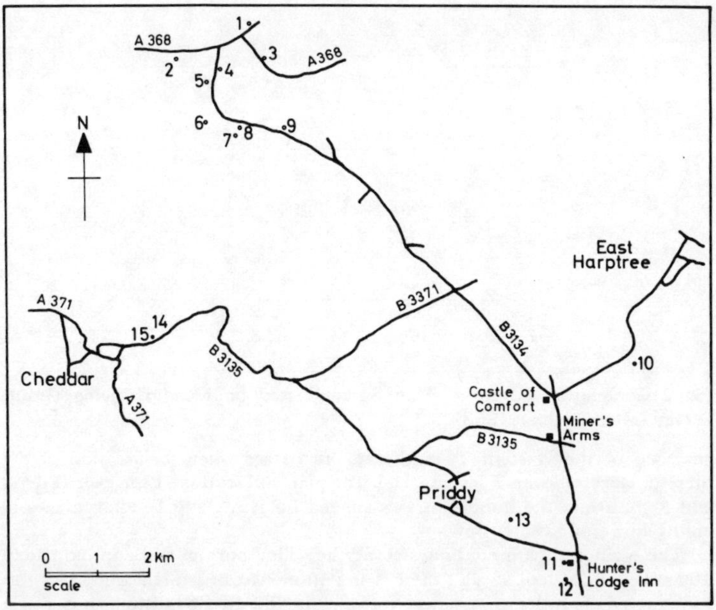

Fig. 1 Locality map for Mendip Hills

was lost. Later excavation in the 1920's from the lower layers in the cave provided a rich late Pleistocene fauna associated with Late Palaeolithic artefacts and human remains (Tratman 1975). The cave is thought to have originated as a rising, the underground tributaries having been captured by the modern risings at Rickford and Langford (Tratman 1963). This statement is supported by the occurrence of large scallops cut into the cave roof and walls. These can be seen in the roof of the cave some 3 m from the entrance and without the need for artificial lighting or caving equipment. These scallops are of the type associated with phreatic flow and have an asymmetric form demonstrating that the flow was out from the cave.

6 Sidcot Swallet (ST 4755 5829) Some 350 m farther up the Combe on the west side is a footpath which follows the line of the Western Twin valley. The lower section of the valley does not contain a stream except under exceptional high flood conditions. There are gravel deposits associated with the lower part of the valley and these were laid down in the exceptional severe floods of July 1968 (Hanwell & Newson 1970). The gravel is sub-angular and is dominantly composed of Old Red Sandstone material. It is thought that this material represents re-deposited head material derived from the flanks of the northern slope of Blackdown.

The entrance to Sidcot Swallet lies some 2–3 m above the footpath to the west; it is immediately above an old manhole cover. The cave has a length of about 220 m and a vertical range of 40 m; renowned for the narrowness of many of its passages, it undoubtedly originated as a former swallet entrance of the Western Twin Stream. In times of high discharge conditions the recent stream disappears through the gravels that infill this section of the valley to depths of about 10 m, and into the solutionally enlarged limestones below.

Goatchurch Cavern is situated some 20 m above the valley floor to the east. This cave has two entrances and a total passage length of about 700 m. The cave is now dry and is best considered as representing a former swallet

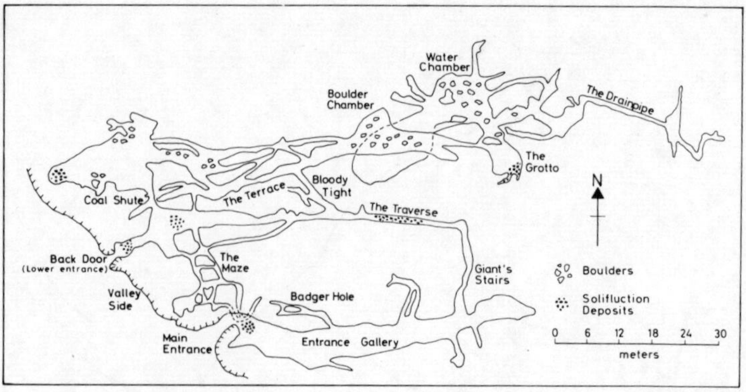

Fig. 2 Goatchurch Cavern – Plan Survey based on Mendip Caving Group Survey (after Tratman 1963)

entrance of the Western Twin Stream at a stage when the incision of the present valley was much less marked. The plan and section of the cave (Figs. 2 and 3) illustrate the important role played by joints and bedding planes in controlling groundwater flow.

The main footpath continues along the valley bottom and approximately 100 m beyond Sidcot Swallet there is an entrance to an artificial adit. This was constructed in order to tap water from the Old Red Sandstone but was a failure in respect of the quantity of water obtained. The geology of the adit is described in detail in Hepworth & Stride (1950). The boundary of the Lower Limestone Shales and the main mass of the Carboniferous limestone is located between the adit entrance and Sidcot Swallet.

At this point the footpath that ascends the western side of the valley should be followed. After about 100 m there is a junction of paths and the left hand path should be followed; this is marked as a footpath on the 1:25 000 map. The path descends the valley side to cross the Western Twin Stream, a good exposure of the Shales can be seen at the stream crossing. The path then climbs up the east side of the Western Twin valley to emerge on a bench-like feature which it follows for about 400 m until it descends again to cross the Eastern Twin Stream.

The bench truncates the Lower Limestone Shales and the main mass of the Old Red Sandstone rises to the south by some 140 m to form the summit of Blackdown. Although the bench is cut across the Shales there are no exposures and the shales are mantled by 10 m or more of Old Red Sandstone debris of periglacial origin.

7 **Eastern Twin Stream** (ST 479 580) The Shales are again exposed at the point where the path fords the Eastern Twin Stream. Some 20 m upstream of this point however, most of the stream flow disappears underground into a thick band of limestone interbedded within the Shales. Prior to 1970 the flow, even at times of low flow, continued over this limestone band but the blocked entrance was dug out in an attempt to find a new cave and subsequently most of the flow takes this route. Under certain flow conditions the water reappears down the valley.

8 **Eastern Twin Swallet** (ST 4796 5814) The route continues either downstream along the stream bed or by following a footpath west of the Eastern Twin. As with the Western Twin the infilling of the valley with coarse sub-angular Old Red Sandstone material is a marked feature; the bedrock is normally only exposed in the stream bed. The wet weather swallet for this

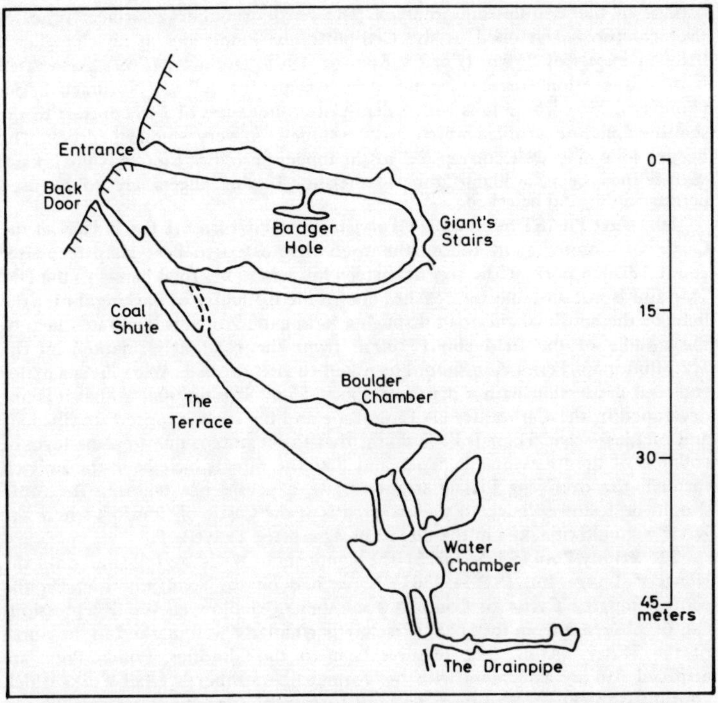

Fig. 3 Goatchurch Cavern – Sectional Elevation E–W. (after Balch 1937)

stream is the Eastern Twin Swallet cave. The entrance passage of the cave can be observed to follow a rift approximately paralleling the strike; the cave has an overall length of about 150 m. The water from the swallet has been traced to both Langford and Rickford risings.

In the July 1968 floods the Eastern Twin Swallet was not capable of taking all the water flowing down the valley. A torrent by-passed the cave to flow down Burrington Combe and laid down a considerable amount of coarse debris and in addition tore up sections of the road surface. The trees below the Eastern Twin Swallet are still partially buried in the gravels laid down by this flood. The details of this flood and its deposition and erosion are described in Hanwell & Newson (1970). (See especially pp.33–4 and their map on Fig. 9).

The footpath joins the main Burrington Combe road about 100 m below Eastern Twin Swallet and the walk back to the parking place takes about 10 minutes. The next section of the itinerary requires motor transport but the use of a coach is not advised at certain of the sites due to problems of parking and turning.

9 **Burrington Combe Head** (ST 489 581) The B 3134 is followed up Burrington Combe; it is possible to park at the top of the combe and to follow a footpath to the north of the road. The soils developed at this location are thin (20–30 cm) and belong to the Lulsgate Series (Findlay 1965). The route continues along the B 3134 and from the road in the locality of ST 496 577 the annular depression which marks the outcrop of the Lower Limestone Shales can be observed; the radio-mast is situated near to the crest of Blackdown on Old Red Sandstone strata.

The B 3134 is followed and near to ST 532 545 good views of the upland surface of the Mendips are obtained. The gentle undulating surface truncates the structures developed in the Carboniferous limestones at an elevation a little in excess of 250 m (Ford & Stanton 1968; Donovan 1969). Associated with this erosion surface are the deep soil profiles of the Nordrach Series (Findlay 1965). These soils with a depth often in excess of 1 m contrast to the shallow Lulsgate profiles which are developed on very similar bedrock. The deeper soils may well correspond to the longer period of time that the upland surface has been available for weathering. If soil augers are to be used permission should be asked.

10 **Wurt Pit** (ST 5588 5391) To visit Wurt Pit turn off the B 3134 at the Castle of Comfort and follow the road signposted to East Harptree, after about 1800 m park at the top of a steep hill where the road bends to the NE. This site is *not* suitable for coaches due to the difficulty of parking. Enter the field to the south of the road through a farm gate. Wurt Pit is situated near to the middle of the field about 100 m from the road; it is marked on the 1:25 000 map. Permission should be asked to visit this site. Wurt Pit is a major enclosed depression with a depth of about 15 m. It is unusual in that it is not developed in the Carboniferous Limestone and the rocks exposed are siliceous and of Liassic age. There is little doubt that the feature is due to some form of collapse of the limestone at depth and that the effects have been transmitted through the overlying Liassic strata to give a subsidence feature. The route should be followed back to the crossroads at the Castle of Comfort where the B 3134 should be taken in the direction signposted to Wells.

11 **Priddy Pool** (ST 547 515) The signposts to Wells are followed until the Hunters' Lodge Inn (ST 549 501) is reached on the road junction. On the route from the Castle of Comfort a number of shallow enclosed depressions can be observed from the road, these are particularly well marked to the north of the Miners' Arms. The features near to the Minories, Priddy Pool, are artificial and are associated with the former St. Cuthbert's Lead Works which continued to extract lead from waste of former mining activity until 1908.

12 **Hunters' Hole** (ST 5492 5005) It is possible to park vehicles in the car park of the Hunters' Lodge Inn public house in order to visit the enclosed depression, Hunters' Hole, which is located some 70 m into the field that lies immediately to the south of the public house. Permission to walk across the field to the depression and to park should be asked at the public house. The depression is about 5 m deep and is noteworthy as a shaft was dug by cavers into the centre of the depression (the entrance is covered by a manhole cover) and although the first 10 m or so are in fractured rock and rubble this becomes a true cave with a vertical depth of a further 40 m which is formed by solutional processes acting on solid, *in situ* rock. This, and similar excavated depressions, demonstrate that at least some of the Mendip depressions are *not* collapse features in the sense that they represent collapse into a large void in the manner found in coalfield areas. Rather such depressions are due to the solutional removal of material where joint and bedding planes are particularly closely spaced and the 'collapse' is a gradual movement infilling these spaces.

13 **Eastwater Cavern** (ST 5387 5062) From the Hunters' Inn take the road to the NW that is signposted to Cheddar. Park beside the road after about 1,000 m opposite a 'dead end' road sign found on the north side of the main road at ST 538 504. Walk along the minor road for about 200 m and turn east through a farm gate just before a green caving hut. Walk some 50 m across the field and the blind valley leading to the entrance of Eastwater Cavern can be seen. A small stream enters the cave which is developed at the bottom of a small limestone cliff composed of massively bedded limestone. Eastwater Cavern has a total passage length of about 1300 m. The valley leading to the cave is deeply incised and no clearly marked dry valley continues on down the

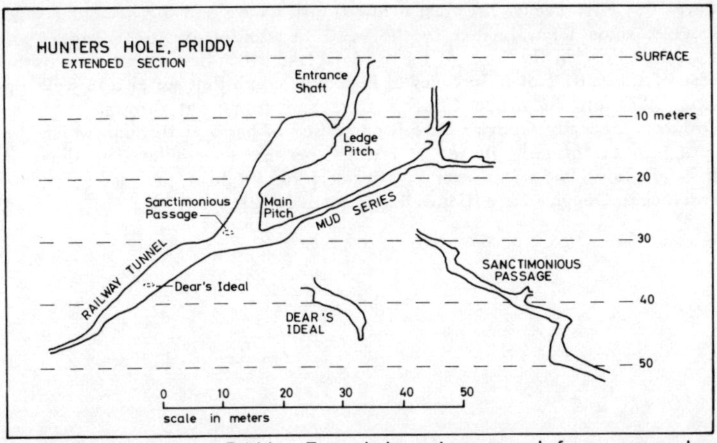

Fig. 4 Hunters' Hole, Priddy. Extended section, traced from survey by B. M. Ellis

valley line. The depth of the incision suggests that the swallet has been located at this point for a considerable period of time; even in times of severe flood little water backs up at the cave entrance and none overflows the depression in which the cave is situated. The water draining into Eastwater Cavern has been traced and reappears at Wookey Hole (Atkinson *et al.* 1967).

14 Cheddar Gorge (Gough's Cave) (ST 4670 5391) The route continues through the village of Priddy and at each junction the signposts to Cheddar are followed. The road joins the B 3135 just beyond the village.

Initially the road beyond the junction with the B 3135 follows the line of a shallow dry valley, but after the junction with the B 3371 the valley quickly becomes incised with progressively higher cliff-like sides as Cheddar Gorge itself is entered. It is not usually practicable to park on the road through the gorge during the summer months. Car parks are located near to the bottom of the gorge at, and around, ST 469 541. It is also possible for coaches to park; it should be noted that in the tourist season coaches are only allowed to travel *down* the gorge. The pressure of tourists during the summer is such that detailed fieldwork is not recommended. However, the grandeur of the gorge and the opportunity to visit a show cave may render a brief visit worthwhile.

The best show cave is Gough's Cave (ST 4670 5391). This is open throughout the year but closes for lunch during the winter. The guided tour takes 20–30 minutes. There is a small museum associated with the cave which is worthy of a visit.

A good viewpoint is obtained from the tower at the top of Jacob's Ladder, the views include the lower section of the gorge, the southern edge of Mendip and the Somerset Levels. Jacob's Ladder is located on the eastern side of the road opposite the Cliff Hotel.

15 Cheddar Gorge (Cooper's Hole and Pride Evans' Hole) (ST 469 541) There are several small, dry caves at varying altitudes on both sides of the gorge and their description and locations are given in detail in Barrington & Stanton (1972). Many of the caves are only reached after rough scrambling on the sides of the gorge which for inexperienced parties can be dangerous. Two caves easily visited are Cooper's Hole which is at road level and Pride Evans' Hole which is situated about 15 m above the road and reached by a short scramble up a path immediately opposite Cooper's Hole.

A number of springs emerge in the section of the gorge below Gough's

Cave. The First Feeder rises just below Gough's Cave and the Second Feeder emerges some 50 m further to the west. In addition up to 16 springs are thought to rise in the floor of the artificial lake which lies immediately to the east of the Cliff Hotel. In times of high flow conditions water also wells up from Skeleton Pit inside Gough's Cave and flows out through the cave entrance. Basically Gough's Cave is composed of passages through which the spring water formerly flowed and which become re-activitated in times of flood. In July 1968 water several metres deep flowed down the road below the entrance to Gough's Cave (Hanwell & Newson 1970).

7
Fossil plants from the Somerset coalfield

K. C. Allen

Objectives

The Somerset Coalfield is for the most part concealed under Mesozoic strata. Coal measure exposures, where they occur are poor, and therefore the principal aim of this excursion is to study the fossil plants found on some of the more recent coal tips.

Special Features

Although all the pits have now closed, permission must be obtained to visit the tips. Applications should be made in writing.

Kilmersdon, Lower Writhlington, Tynings Tip and New Rock Tip: The Area Secretary, South Wales Area, National Coal Board, Coal House, TyGlas Avenue, Llanishen, Cardiff.

Marsh Lane tip: The Duchy of Cornwall Office, 5 Edward Street, Bath.

Old Mills tip: Clapton Building Supplies Ltd., Old Mills, Paulton, Avon. In the case of Old Mills tip state that the company will be held in no way responsible for any injuries that might occur.

Maps

Ordnance Survey	One-Inch Sheet 166 (Frome)
	1:50 000 Sheets 172 (Bristol & Bath)
	183 (Yeovil & Frome)
	1:25 000 Sheets ST 65, ST 66, ST 75
Geological Survey	One-Inch Bristol District

Main References

Crookall (1925a & b); Moore & Trueman (1937); Chaloner & Collinson (1975).

Outline Geology

Coal has been mined in the Somerset Coalfield since Roman times. However, it was not until the middle of the seventeenth century that it became a notable industry. Coal mining reached its peak during the first quarter of the twentieth century, after which it gradually declined. At the time of Nationalisation in 1947, there were only twelve working mines; by 1958 the number was reduced to five, and after 1968 to two, namely Kilmersdon and Lower Writhlington. These two, which were in later years administered as one, finally closed in 1973.

The Somerset Coalfield, apart from two isolated areas to the west, is an oval-shaped basin, affected by both N–S and E–W faults and overthrusts. The latter are the most important, and two of them, the Farmborough Compression Belt, and the Southern Overthrust, separate the basin into three parts: the Pensford Basin to the north, the Radstock Basin in the centre, and the Nettlebridge Area to the South. The collieries in the Pensford and Radstock Basins worked mainly the Upper Coal Measures, those in the Nettlebridge Area mostly the Middle and Lower Coal Measures. That part of the Coal Measures section relevant to this excursion is shown in Fig. 2. For stratigraphical nomenclature, Curtis *et al.* (1973, p. 20) has been followed. In the Radstock Basin, the coal seams from the Radstock Group being the highest, were the first to be worked, and many of the earlier records of fossil

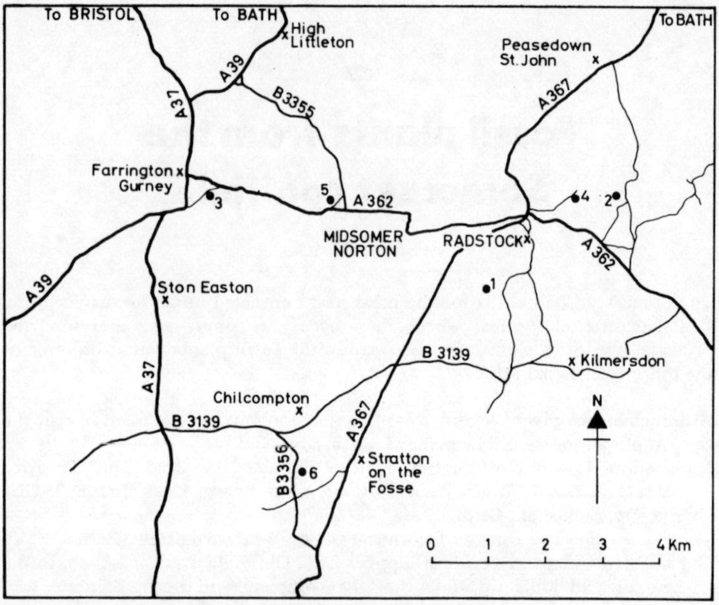

Fig. 1 Map showing the location of the tips described

plants from the Somerset Coalfield came from the shales above these seams (see Kidston 1887, and Crookall 1925a, b). However, most of these have long been exhausted, and the last colliery to work coal seams in the Radstock Group was Camerton, which closed in 1950. In the Pensford Basin however, Pensford Colliery worked the Pensford No. 2 and No. 3 seams up until its closure in 1958, and these are thought to correlate with lower seams from the Radstock Group in the Radstock Basin (Kellaway 1970, p. 1047). Pensford Tip has produced good fossil plants in the past, but access is no longer possible. With the Radstock Group being worked out in the Radstock Basin, the shafts were sunk through into the Farrington Group, and most of the later collections have come from the shales above nine seams in this group (Itinerary 1, 2, 4, 5). From the Mangotsfield Group, only the Rudge has been mined commercially (Itinerary 3). In the Nettlebridge Valley, the lower seams from the Downend Group, and the Middle and Lower Coal Measure seams were worked in the past (Crookall 1925; Moore & Trueman 1937), but collecting in this area is now very poor, and only New Rock Tip (Itinerary 6) is listed.

It is usually only the roof shales and silty mudstones from just above the coal seams, which are removed to the tips. The fossil flora found will therefore not represent the complete coal measure flora growing in the region, but probably only those plants growing in the lower parts of the swamp forest. Unfortunately, on weathering, the roof shales tend to fragment very easily, so that the large rock specimens are found only at the most recent tips, namely Kilmersdon and Lower Writhlington.

A good pocket book for identifying Coal Measure plants, both by photographs and with a key is *Steinkohlenpflanzen* by Gothan & Remy (1955). Other useful publications for identification, include the palaeontology memoirs of the Geological Survey of Great Britain, Vol. 2 (Kidston 1923–5) and vol. 4, parts 1–6 (Crookall 1955–70), and *Coal Measure Plants* by

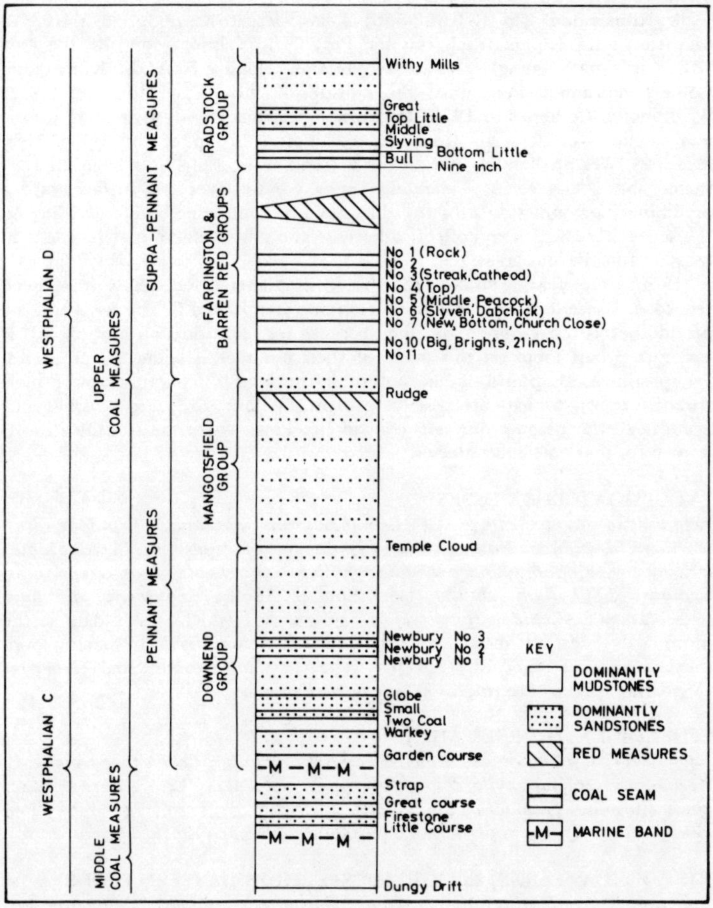

Fig. 2 A generalised succession through that part of the Coal Measures in the Somerset Coalfield referred to in the text

Crookall (1929). Chaloner & Collinson (1975) have just produced an illustrated key to the commoner British Upper Carboniferous plant compression fossils which will be an especially useful field guide for non-palaeobotanists.

Itinerary

Six coal tips are described (see Fig. 1), but it is not essential that they should all be visited, or that they should necessarily be visited in the listed order; this will depend on time available, and the direction of travel. The best collecting is at Kilmersdon (**1**) and Lower Writhlington (**2**), but Marsh Lane (**3**) and Tyning (**4**) provide small, easily accessible collecting areas. Old Mills (**5**) and New Rock (**6**) are now rather poor. A detailed list of plants collected over the last few years is given for Kilmersdon, because there are no published fossil lists from the shales above the No. 10 seam in the Farrington Group, worked since 1951 at Kilmersdon, and 1960 at Lower Writhlington. A list of plants recently recorded at Marsh Lane is also given but from the other tips only certain plants or groups are mentioned.

1 Kilmersdon Tip (ST 682 536) Leave Radstock on road south to Charlton; after 1.5 km reach Haydon Post Office; almost opposite the Post Office is a path going westward to the tips. Tipping from the Kilmersdon Colliery continued here until the closure of the Kilmersdon and Lower Writhlington Collieries in 1973. During the previous twenty-two years mining was in the No. 10 seam from the Farrington Group of the Upper Coal Measures (Westphalian D), and all the material now available is from the roof shales above this seam. Collecting varies greatly over the surface, and a preliminary reconnaissance of the tip should be completed before deciding on the areas from which to collect, otherwise too much time might be spent in sparsely fossiliferous areas.

During the years 1968–1973, the fossil plants listed below have been recorded. Those that are particularly common are marked **C**. The ferns and the pteridosperms are grouped together because for identification purposes, it is easier to group them on the shape of their pinnules as being Sphenopterid, Pecopterid, Alethopterid or Neuropterid, rather that to separate them on their fructifications, which are rarely present. Within the large form-genus *Pecopteris*, the organ-genus is shown in brackets, where the fructification is known for that particular species.

LYCOPSIDA (CLUB MOSSES)
Lepidodendron aculeatum **C**, *Lepidodendron ophiurus, Lepidodendron wortheni, Lepidodendron acutum, Lepidodendron rimosum, Lepidophloios laricinus, Lepiodophylloides (Lepidophyllum)* sp., *Lepidostrobus variabilis, Lepidostrobophyllum alatum* (see Boulter 1968), *Lepidostrobophyllum lanceolatum* **C**, *Lepidocarpon major, Sigillaria* sp., which is abundant in the more coaly beds, but usually decorticated and unidentifiable to specific level, most being of the Rhytidolepis type. *Sigillariostrobus* sp., *Cyperites (Sigillariophyllum) bicarinatus* **C**, *Stigmaria ficoides.*

SPHENOPSIDA (HORSE TAILS)
Calamites cisti, Calamites suckowi, Calamites undulatus, Calamites carinatus **C**, *Annularia sphenophylloides, Annularia stellata* **C**, *Asterophyllites equisetiformis, Pinnularia columnaris, Calamostachys* sp. *Sphenophyllum emarginatum* **C**, *Sphenophyllum cuneifolium, Sphenophyllostachys* sp.

PTEROPSIDA (FERNS) and PTERIDOSPERMOPSIDA (SEED FERNS)
Sphenotperids: *Mariopteris nervosa, Mariopteris plumosa, Sphenopteris macilenta.*
Pecopterids: *Pecopteris (Dicksonites) pluckeneti, Pecopteris (Asterotheca) miltoni, Pecopteris (Asterotheca) cyathea, Pecopteris (Asterotheca) arborescens, Pecopteris (Acitheca) polymorpha* **C**, *Pecopteris (Ptychocarpus) unitus, Corynepteris sternbergi.*
Alethoterids: *Alethopteris serli* **C**, *Alethopteris ambigua, Alethopteris grandinioides.*
Neuropterids: *Neuropteris ovata* forma *typica, Neuropteris ovata* forma *flexuosa, Neuropteris heterophylla, Neuropteris macrophylla, Neuropteris scheuchzeri* **C**, cyclopterid pinnules.
Seeds: *Trigonocarpus* sp.

CORDAITALES
Cordaites sp., *Samaropsis* sp.

During the same period, few animal fossils have been found apart from occasional lamellibranchs, though during the last three visits Bristol University students have found both *Euproops kilmersdonensis* and *Eophrynus jugatus,* two arthropods recorded by Ambrose & Romano (1972).

2 **Lower Writhlington Tip** (ST 703 552) Leave Radstock on A 362 and at Upper Writhlington turn north along a minor road to Lower Writhlington. Access to the tip is from the south side over the bridge crossing the Wellow Brook. For thirteen years prior to the pits' closure in 1973, Lower Writhlington like Kilmersdon worked the No. 10 (Big, Brights, 21 inch) seam from the Farrington Group. Before this, the Farrington No. 5 and No. 6 seams were worked. However, it is possible that all the collecting will be from the shales above the No. 10 seam. The last seams from the Radstock Group, were worked here in 1924.

Collecting from this tip is as good or perhaps even better at present than at Kilmersdon. The species list is the same, and will not be repeated here, but the abundance of certain species is different, and both tips should be visited. At present, Kilmersdon has more lycopsids, whilst at Lower Writhlington, pteridosperms and ferns are particularly abundant. Collecting is good over most of the tip, but is best on the western side. From the tip, there are good views of Tyning Tip (ST 695 553, see locality 4) and the lower Braysdown tip (ST 704 552) now grass and tree covered, and no longer worth a visit.

3 **Marsh Lane Tip** (ST 632 553) A small, grass-covered tip on the east side of the minor road connecting the A 362 and A 37, 0.5 km southeast of Farrington Gurney. Marsh Lane Colliery closed in 1949, and few traces remain apart from the tip. The No. 10 seam from the Farrington Group, was worked here for many years, but during the last three years of its life the Rudge was mined; the only seam to be mined commercially from the Mangotsfield Group of the Upper Coal Measures in the Somerset Coalfield.

This is a good trip to start at if coming from Bristol; for although the shales are exceptionally fragmentary, the tip is small, easily accessible, and quite fossiliferous. The best material is along the west side of the tip nearest the road.

Fossil plants recorded here between 1970–1973 include *Lepidodendron* sp., *Lepidostrobophyllum lanceolatum, Cyperites bicarinatus, Annularia sphenophylloides, Annularia stellata, Asterophyllites equisetiformis, Sphenophyllum emarginatum, Sphenophyllum majus, Mariopteris nervosa, Sphenopteris macilenta, Pecopteris (Asterotheca) miltoni, Pecopteris (Asterotheca) cyathea, Alethopteris serli, Alethopteris lonchitica, Neuropteris scheuchzeri, Neuropteris rarinervis, Neuropteris* sp., cyclopterid pinnules, *Trigonocarpus* sp.

It is likely that most of the collecting is from the Rudge (see Fig. 2), probably equivalent to a Lower Westphalian D age. The flora, though not so diverse, is similar to that recorded at Kilmersdon and Lower Writhlington, though *Sphenophyllum majus,* a plant more characteristic of Westphalian C, occurs here. *Asterophyllites equisetiformis* is more frequent than at Kilmersdon.

4 **Tyning Tip** (ST 697 553) Leave Radstock on A 367 going north and just beyond the railway crossing turn right into Waterloo Road. Early plant collections from this tip may come from Tyning, Middle or Ludlows Collieries, all of which tipped here. Tyning tip was noted for the discovery of the large dragonfly-like *Boltonites radstockensis.* The older part of the tip to the north side of Waterloo Road, is now grass-covered and wooded. Tipping continued to the south side until Ludlows Colliery closed in 1954. Much of this part of the tip is also wooded, particularly the southern part adjacent to the railway, but a small area at ST 696 552 opposite the most easterly conical tip, provides good plant collecting. This was the last material to be tipped and comes from the roof shales above No. 11 seam, the lowest in the Farrington Group (see Fig. 2). Particularly good here, are species of *Neuropteris;* though

Alethopterid, Pecopterid and Sphenopterid pinnules are also present. Other fossils more common here than at Kilmersdon and Lower Writhlington are *Stigmaria ficoides* and *Cordaites* sp. The flora is typical of Westphalian D.

5 **Old Mills Tip** (ST 654 551) This tip, on the north side of the A 362, 0.4 km west of the crossing with B 3355 at Midsomer Norton, has produced many of the earlier plant fossils (Kidston 1887; Crookall 1925a, 1925b) now provides rather poor collecting, and should only be visited if time allows. Old Mills Colliery closed in 1966 having latterly worked the No. 5 (Middle) and No. 7 (Bottom) seams from the Farrington Group. The rock material is mostly very small, but the larger fragments yield fossils common in Westphalian D, e.g. *Alethopteris serli Neuropteris scheuchzeri, Annularia stellata* and *Calamites carinatus. Cordaites* and *Sigillaria* are common in the more carbonaceous shales. The best collecting is to the north of the 50 m high dirt tip.

6 **New Rock Tip** (ST 649 506) Take the A 367 south from Radstock to Stratton on the Fosse. The tip is on the east side of the B 3356, adjacent to Downside Abbey and 1 km west of Stratton on the Fosse. The colliery is now in private hands, but the tip is still under the National Coal Board. New Rock Colliery was probably sunk in 1819 (Down & Warrington 1971, p. 251), and worked the lower seams from the Downend Group of the Upper Coal Measures, and the upper seams from the Middle Coal Measures. It is included in the itinerary because it represents the best collecting now available from the lower part of the succession. Fossil plant collecting is however poor compared with Upper Coal Measure tips in the Radstock Basin, and is only recommended if there is time available. The last and lowest seam to be worked at New Rock was the Great Course (see Fig. 2), and most of the collecting is from the shales and sandstones above this seam. The tip slopes have now been planted with conifers, but the larger shale and sandstone fragments contain *Stigmaria, Calamites* and occasional neuropterid and pecopterid pinnules.

Acknowledgments

Thanks are given by the author to the National Coal Board at Kilmersdon Colliery who supplied details of coal seams worked at various collieries.

8
The geology of Cattybrook brick pit, Almondsbury

P. L. Hancock and B. P. J. Williams

Objectives

The purpose of this excursion is to examine the details of the geology of an inlier of deformed Coal Measure rocks surrounded by an unconformable cover of gently tilted Triassic sediments. Because the Coal Measure rocks are highly deformed special emphasis is given to tectonic structures.

Special Features

Access to the pit (ST 593 834) from Bristol is via Cribbs Causeway (A 4018) and then the B 4055 to Almondsbury for about 2.5 km from the motorway intersection (Fig. 1). Although clay has been worked since 1865, the present pit, open since 1929, is the only one in which the geology is well-exposed. It is sited immediately NE of the brickworks.

Permission to enter should be obtained from: The Manager, Cattybrook Brick Co. Ltd., Almondsbury, Bristol. Because the pit is being actively worked the geology available for inspection varies with the state of the excavations, the construction of roadways, the tipping of spoil, and the flooding of some shallow workings. Wellington boots are advisable during wet periods of the year.

Maps

Ordnance Survey	1:50 000 Sheet 172 (Bristol & Bath)
	1:25 000 Sheet ST 58
IGS	1:63 360 Bristol District (Special Sheet)

Outline Geology

Carboniferous The Upper Carboniferous rocks are situated on the eastern margin of the Severn Coalfield which is concealed except for a few small outcrops such as that at Cattybrook. Welch & Trotter (1961) report that coal was worked at Cattybrook in 1756, and that the presence of the concealed basin was predicted by Buckland & Conybeare (1824).

The fossil flora indicates that the rocks belong to the *modiolaris* zone of the Lower Coal Series (Moore & Trueman 1942). Structural complexity makes it difficult to establish the complete Coal Measure succession. It is mainly a mudstone sequence with subordinate sandstones concentrated at two principal horizons. There are also thin beds of intraformational conglomerate, anthracitic coal, fireclay and ironstone. Concretions of clay ironstone are abundant in some mudstones. Excellently preserved plant fossils may be collected from some of the less deformed mudstones.

The rocks at Cattybrook Brick Pit are deformed in a complex zone beneath the Ridgeway Fault: a structure of Variscan age which near the pit thrusts the Carboniferous Limestone, on the west margin of the Bristol Coalfield, NW over the measures of the Severn Coalfield. The adjacent inlier of Carboniferous Limestone at Knole Park is separated from the Cattybrook Coal Measures both by the fault and by the Keuper cover.

The disturbance of the Coal Measure rocks was recognised by Smith & Reynolds (1929), and a recent resurvey has confirmed Prestwick's observation

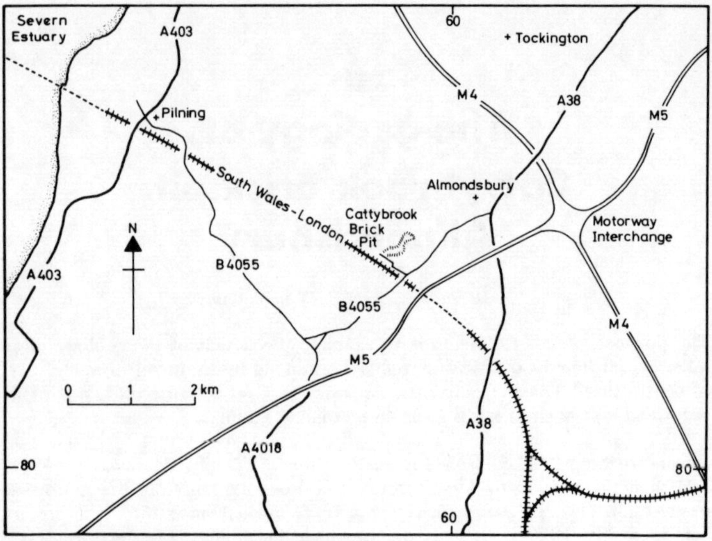

Fig. 1 Location of Cattybrook Brick Pit

made in 1871 that the principal structure is an anticline. The anticline is an asymmetric structure plunging SW and verging NW (Fig. 2). In its core and on its limbs there are parasitic folds which are mainly congruent and plunging to the SW. The steep limbs of both large and small folds are nearly vertical or overturned by up to 30°. Most axial planes dip steeply SE, that is most of the folds verge NW, a sense of overturning which is in accord with the direction of thrusting on the Ridgeway Fault. The majority of the pit has been excavated in the steep or overturned NW limb and core of the principal anticline.

Reverse strike-faults are probably associated with many of the folds, but it is difficult to trace individual examples for more than a few metres. Some cross-faults, along which there has been a component of strike-slip, displace beds on the steep limb of the principal anticline. Minor structures associated with the folds include a spaced cleavage in some mudstones, ubiquitous polished and slickensided shear surfaces, joints, and excellent examples of the types of minor faults which Norris (1958) has called contraction and extension faults.

These terms were introduced to describe small faults which either shorten or elongate layers and which generally form conjugate sets orientated symmetrically about bedding surfaces (Fig. 3). Contraction and extension faults are analogous to reverse and normal faults respectively when bedding, rather than the present-day horizontal, is the plane of reference. Contraction faults generally subtend angles of less than 45° with bedding, while extension faults usually make angles of greater than 45° with bedding. Many of the faults intersect bedding surfaces parallel to an adjacent fold hinge line, thus where the faults are closely-spaced, a *b* lineation parallel to fold axes results. Relative to the presumed sense of inter-stratal slip during folding, the slip on an individual fault may be described as either antithetic or synthetic (Fig. 3).

At Cattybrook conjugate sets of closely-spaced extension faults at about 60° to the bedding are well-developed in some steeply inclined sandstones. There are only a few isolated examples of contraction faults.

The relative abundance of extension faults in the steeply inclined

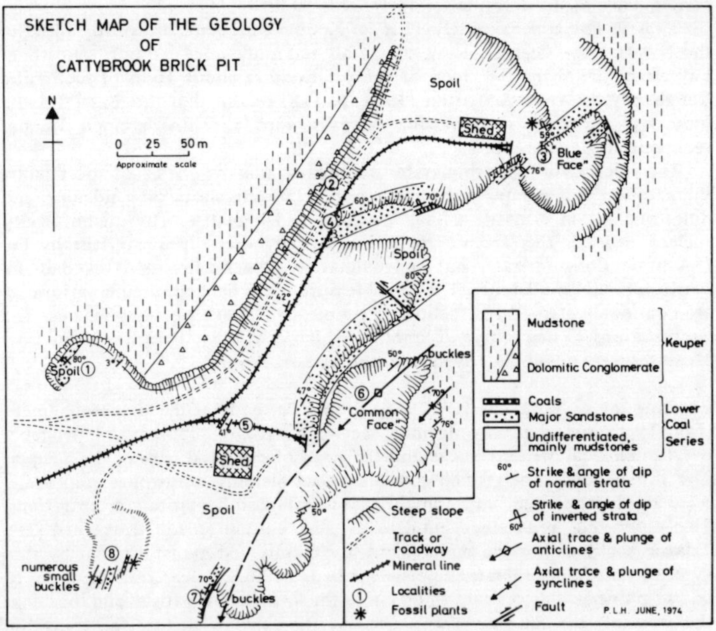

Fig. 2 Sketch map of the geology of Cattybrook Brick Pit

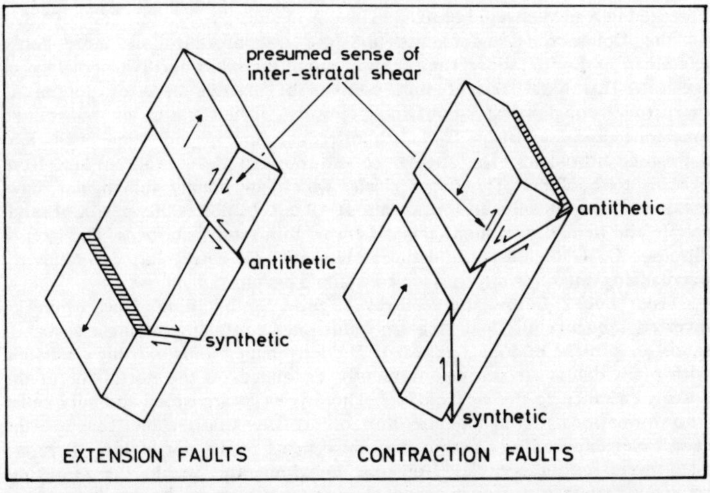

Fig. 3 The geometry of extension and contraction faults relative to layering

sandstones is probably a reflection of the stretching which occurred in them during the later stages of folding. Most of the relatively ductile mudstones probably responded to this stretching by flow, but the sandstones being more brittle, failed on conjugate shear surfaces. Some of the thin beds of ironstone in the abandoned pit to the east are reported to be pulled apart as a consequence of the deformation (Reynolds 1929).

Triassic The approximately 15 m of Keuper sediments are gently tilted to the NW at a few degrees. They are mainly red mudstones and buff sandstones but along the main NW face of the pit there is about 10 m of Dolomitic Conglomerate. Welch & Trotter (1961, p. 118) record that this Basal Triassic unit was formerly seen passing laterally into the finer-grained Keuper sediments at Cattybrook.

The Dolomitic Conglomerate comprises massive beds of polymictic sharpstone conglomerate containing poorly sorted clasts of sandstone and limestone set in a matrix of sand, calcite and hematite. The unconformity surface beneath the Triassic rocks is irregular where it is overlain by the Dolomitic Conglomerate, but approximately planar where it is overlain by sandstones and mudstones. The Coal Measure rocks beneath the unconformity are commonly stained red, yellow or purple to a depth of a few metres. The encrustations of hematite and limonite which coat many fractures in the Coal Measures were possibly derived from the Triassic cover.

Itinerary Entering the pit from the NW a stop should be made at the first embayment (Loc. 1) to see the unconformity between steeply dipping and probably overturned Coal Measure rocks and the nearly horizontal beds of the Keuper. The thin-bedded, fine-grained quartzitic Coal Measure sandstones are parallel – or ripple-laminated, and contain plant casts and hematite concentrations. The contorted, red-stained mudstones also contain plant fragments. The Triassic rocks within the embayment are mainly red mudstones cut by thin gypsum veins. Subordinate impersistent beds of buff, calcareous, medium- to coarse-grained sandstones up to 60 cm in thickness occur within, and locally at the base of, the mudstone sequence. At the time of writing the transition between the mudstones and sandstones in the embayment, and the Dolomitic Conglomerate, forming the prominent ledge at its southeastern extremity, is obscured in a grass-covered bluff.

The Dolomitic Conglomerate and the unconformity are more easily examined at Loc. 2 above the roadway along the main northwestern face of the pit. This basal Triassic unit consists of massive beds of polymictic sharpstone conglomerate containing poorly sorted clasts of red-stained limestone and sandstone of Carboniferous aspect, and some Old Red Sandstone lithologies. The angular to sub-rounded clasts range in size from pebbles to boulders. The larger clasts, which are mainly sub-angular, have maximum long axis dimensions of about 40 cm. In places the matrix of sand, calcite and hematite is dominant and forms thin sandy interbeds. The steeply dipping Coal Measure mudstones beneath the irregular, prominently overhanging unconformity surface are stained purple and yellow.

From Loc. 2 follow the roadway to Loc. 3, the 'Blue' face, where an inverted sequence of Coal Measure mudstones containing a single group of sandstones in the middle is exposed. The lens-shaped units of lithic sandstone with plant debris are most conveniently examined on the north side of the narrow entrance to the worked face. There is a clay ironstone and mud pellet (intraformational) conglomerate horizon at the stratigraphic base of the sandstones exposed just within the embayment on its north side. There are also several subsidiary clay ironstone conglomerates within the sandstone sequence. These clay ironstone conglomerates, which are known to occur at the base of Coal Measure sandstone units, indicate that the sequence is overturned.

Thin lenticular coal seams occur within the mudstones which yield abundant and well-preserved plant fossils from the beds immediately and stratigraphically above the sandstones. Excellent specimens of *Calamites* stems, *Asterophyllites* (*Calamites* leaves), *Pinnularia* (*Calamites* roots),

Neuropteris, Cyclopteris (basal leaves of *Neuropteris*) *Mariopteris* and *Alethopteris* may be collected.

An example of a small contraction fault, displacing a bed of sandstone for about 30 cm, occurs half-way up the bluff of sandstones on the north side of the entrance to the 'Blue' face. Although the fault is inaccessible it can be seen to be a strike-fault subtending an angle of less than 45° with the bedding.

Looking WSW when cautiously approaching Loc. 4 along the mineral line it is possible to see high in the opposing cliff overturned mudstones and coals beneath a thick sandstone sequence. These sandstones are best examined at Loc. 4, a narrow cutting on the mineral line, where a basal clay ironstone conglomerate indicates that the steeply northwest dipping beds are not overturned. In detail the sequence comprises beds of lithic sandstone totalling about 15 m in thickness with interbeds of thin grey mudstones near the top. Some of the sandstones exhibit obscure large-scale trough cross-bedding and some are internally parallel-laminated.

On the eastern side of the cutting two major west-facing sandstone bedding surfaces are disrupted by conjugate sets of closely-spaced extension faults at a high angle to the bedding. The faults form a conspicuous lineation plunging gently NE on the bedding. This lineation is especially noticeable because it results not only from the intersects of the faults on the bedding, but also from the steps arising out of the repeated displacement of the bedding surfaces. A third set of less abundant extension faults forms a lineation which plunges gently SW. Antithetic extension faults outnumber synthetic extension faults. In the west wall of the cutting some of the fault surfaces are polished and lineated by slickensides. The extension faults are absent or poorly represented in the mudstone interbeds, and they are noticeably less abundant in the lower parts of sandstone units. Many iron-stained dip joints cut the extension faults.

About 10 m NE of the cutting a minor flexure zone with steep hinge lines plunging approximately down the dip of the sandstones deflects the strike of the beds in a dextral manner.

Loc. 5 is reached by following the mineral line SW to its junction with the line to the 'Common' face and then following that line through a low-sided cutting. In this cutting through the Coal Measures there is a west dipping sequence of mudstones with lithic and quartzitic sandstones near its base. A thin deformed coal seam is exposed on the south side of the cutting near to the junction of the lines. A reticulate spaced cleavage is developed in some of the mudstones.

Loc. 6, the 'Common' face is another part of the pit which is being worked. In it a thick sequence of Coal Measure mudstones is exposed in the core and on the flanks of the principal anticline. Measurements from the limbs of this fold show that it plunges gently SW and verges to the NW. In the mudstones there are, in addition to clay ironstone concretions, subordinate thin beds of parallel and ripple-laminated, fine-grained sandstone, coal, fireclay and oolitic ironstone. Some of the mudstones yield excellent plant fossils. Depending upon the state of the workings, minor buckles in the core of the fold may be visible. Nodules containing hematite, goethite, celestite and calcite may be collected from spoil on the floor of the 'Common' face.

At Loc. 7 the Coal Measure sequence comprises grey folded and thrusted mudstones, which are cleaved in places, coaly layers up to 3 cm in thickness, ironstone concretion horizons, and thin beds of fine-grained, ripple-laminated sandstones. Several of the steeply dipping sandstones are disrupted by closely-spaced conjugate extension faults forming a conspicuous lineation where they intersect bedding surfaces. The faults are more frequent in the thin sandstones at this locality than they are in the thicker sandstones of Locality 4. Again antithetic faults outnumber synthetic faults.

A steeply plunging flexure zone, similar to that at Locality 4, deflects the strike of the sandstones and the conspicuous lineation formed by the intersects of the extension faults on the bedding. The sense of shear on the zone is sinistral and the amount of displacement about 15 cm.

A final stop should be made at Loc. 8, a shallow excavation which is sometimes flooded, where a sequence of folded mudstones and sandstones with subordinate clay ironstones and coals is exposed. The sandstones which crop out in the central part of the section are quartzitic, fine-grained and ripple-laminated. The sequence is deformed by at least ten minor west verging folds of which the axial traces of only three are shown on Fig. 2. The steep limbs of some of the folds are overturned by a few degrees and several of the anticlines are relatively tight cuspate folds.

9
The Mesozoic strata of the Mendip Hills

R. J. G. Savage

Objectives

The excursion is designed to demonstrate the succession of Mesozoic strata around the Mendip Hills. These range from the Triassic Dolomitic Conglomerate and Keuper Marls through Rhaetic and Lias to Inferior and Great Oolite in the Jurassic. The normal facies are seen some distance away from the Mendips and littoral facies examined around the foothills. Additionally some Mesozoic fissure infillings are examined; these are famous for their terrestrial vertebrate faunas.

Special Features

Permission is required to visit a number of localities and this may be obtained as follows:

Brockley Combe quarry: Mrs. Cook, Downside House Farm, West Town, Backwell. Lulsgate 2353.

Wurt Pit: Mr. W. J. Masters, Ebbor Farm, Wookey Hole. Wells 73067.

Emborough quarries: Mr. M. Blatchford, Ralph Blatchford & Co. Ltd., St. Barnabas Works, West Road, Midsomer Norton, Bath, Somerset. Midsomer Norton 412281.

Doulting quarry: The Manager, Dorset Limestones Ltd., Doulting Quarry, Doulting, Shepton Mallet, Somerset.

Merehead quarry: The Manager, Merehead Quarry, Foster Yeoman Quarries Ltd., Torr Works, Cranmore, Shepton Mallet, Somerset. Shepton Mallet 3311.

Holwell quarries: The Manager, English China Clays, Coleman's Quarry, Holwell, Frome, Somerset. Nunney 401.

Maps

Ordnance Survey	One-Inch Sheets 165 (Weston-super-Mare) and 166 (Frome)
	1:50 000 Sheets 182 (Weston-super-Mare & Bridgwater)
	183 (Yeovil & Frome)
	1:25 000 Sheets ST 46, ST 55, ST 56, ST 64, ST 65
	ST 74
Geological Survey	One-Inch Bristol District Sheet;
	Sheets 280 (Wells), 281 (Frome)

Main References

Green & Welch (1965); Robinson (1957).

Outline Geology

The history of the Mendip hills during Carboniferous and Permian times is summarised in Excursion 5, Outline Geology (p. 55). During Triassic times the climate was tropical, hot with alternating wet and dry seasons: the hills were being eroded subaerially and piedmont facies accumulated around the foothills – Carboniferous pebbles, cobbles and boulders brought down wadis in the rainy season, perhaps monsoons; these now form the Dolomitic Conglomerate (loc. 12). The conglomerate matrix is often rich in hematite and geodes of silica and calcite. The occurrence of hydro-thermal mineralisation with lead-zinc ores in the Mendip hills was largely controlled by the anticlinal

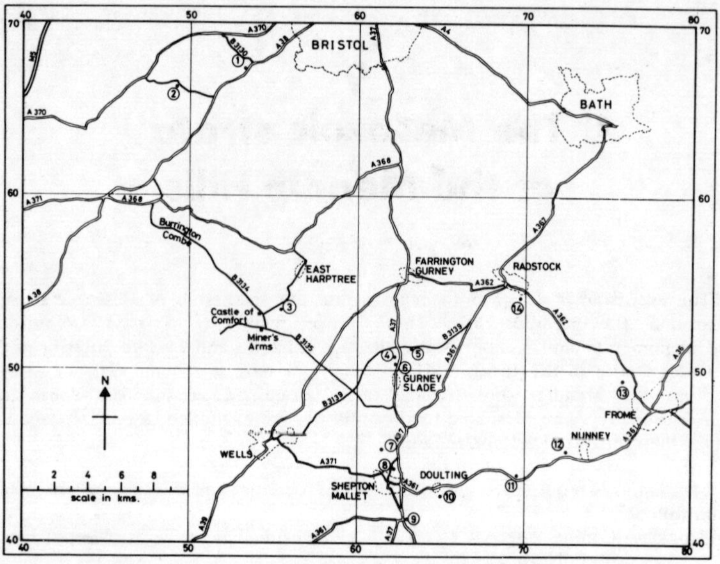

Fig. 1 Locality map

structures and permeability of the beds: it is considered to be at least in part of Keuper age. Green (1958) has given an excellent account of the mineralisation in the central Mendips to which the reader is referred for details.

On the plateau sink holes appeared and grikes developed along joint openings. Into them was washed fine sediment, now trapped in some of the fissure infillings along with relicts of the plant and animal life on the hills. In other instances these sediments found their way into cave systems which had developed earlier under phreatic conditions.

Whittaker (1973) has shown that the Central Somerset Basin is a structural basin, continuous with the Bristol Channel to the west and probably continuous with the Wessex Basin to the east. In this intermontane basin thick Mesozoic strata accumulated. The W–E trend is of Variscan origin, but movements continued through the Mesozoic. Sediments are thickest toward its centre and thin against the margins, with often littoral facies developed around the Mendip hills. The Rhaetic provides a good datum for deformation as it is always thin and easily recognised. In the basin, at Brent Knoll, Rhaetic deposits occur at -411 m OD; on the Mendips they rest at +244 m OD; this accounts for a relative downwarping of 650 m since Rhaetic times. The steep sides and broad flat bottom to the trough suggest a fault bounded graben structure.

During Rhaetic times the sea transgressed toward the area and brackish lagoons formed into the shallows; around the hills sandy facies developed and occasional conglomerates (loc. 13) formed. During Liassic times the sea transgressed further into the area swamping the lagoons (see Fig. 2). Deep water conditions prevailed south of the Mendip hills (loc. 9) while to the north towards Bristol shallow seas gave rise to shelly limestones (loc. 14). Around the foothills littoral detrital limestones of Carboniferous origin accumulated (loc. 8). North of the Mendips, Broadfield Down appears to have been a shoal during early Liassic times with again a littoral facies developed (loc. 2).

			STAGES
UPPER JURASSIC		Cornbrash	
GREAT OOLITE		Forest Marble Upper Fuller's Earth Fuller's Earth Rock Lower Fuller's Earth	BATHONIAN
INFERIOR OOLITE	Up	Anabacia Limestones Doulting Stone & Conglomerate	BAJOCIAN
	Mid	(absent)	
	Low	(absent)	AALENIAN
	Up	Clays or sands Junction Bed	TOARCIAN
	Mid	(absent except on Dundry Hill)	
	Low	Clays; lower part passing into ironshot, shelly limestones in Radstock area	PLIENSBACHIAN
		Blue Lias (Downside Stone locally)	SINEMURIAN HETTANGIAN
RHAETIC		White Lias or Langport Beds Cotham Beds Westbury Beds	RHAETIAN
TRIASSIC		Keuper Marls Dolomitic Conglomerate	

Stratigraphic Table of units and stages used in text

Shallowing of the seas and subaerial exposure in places has resulted in the absence of Middle and almost all of the Upper Lias, early and middle Inferior Oolite sediments from the area of the Mendips. The next sediments preserved are those of the late Inferior Oolite laid down in the wake of the Upper Bajocian transgression which swept eventually across much if not all the Mendip island. These sediments lie with marked non-sequence on earlier Jurassic strata and often overstep onto the Carboniferous (loc. 12 & 13). It was during this long interval of about 15 million years between the early Liassic and late Inferior Oolite times that much interest in the vertebrates centres. During this time span while we have almost no normal sedimentation, the subaerial erosion products were being trapped in fissures and cave systems, mostly capped by normal marine sediments of Upper Inferior Oolite age. Into the fissures were washed the soft relatively unconsolidated Rhaeto-Liassic sediment, the more heavily eroded products of the Carboniferous limestones and occasionally pebbles of Devonian and Silurian facies. The streams also carried along the remains of contemporary animals, mostly reptiles, to be deposited juxtaposed with remanié vertebrates, mainly fish from the earlier strata (loc. 7 & 12), thus making the precise age of the infillings very difficult to determine. Robinson (1957) argued that these fissures were neptunian dykes infilled while submerged under the sea in Rhaeto-Liassic or early Inferior Oolite times and that the marine and terrestrial faunas could thus be at least in part contemporary. A further period of hydrothermal silicification affected the Harptree area in late or post Inferior Oolite times, giving rise to the silicified Harptree beds (loc. 2). The youngest Mesozoic beds around

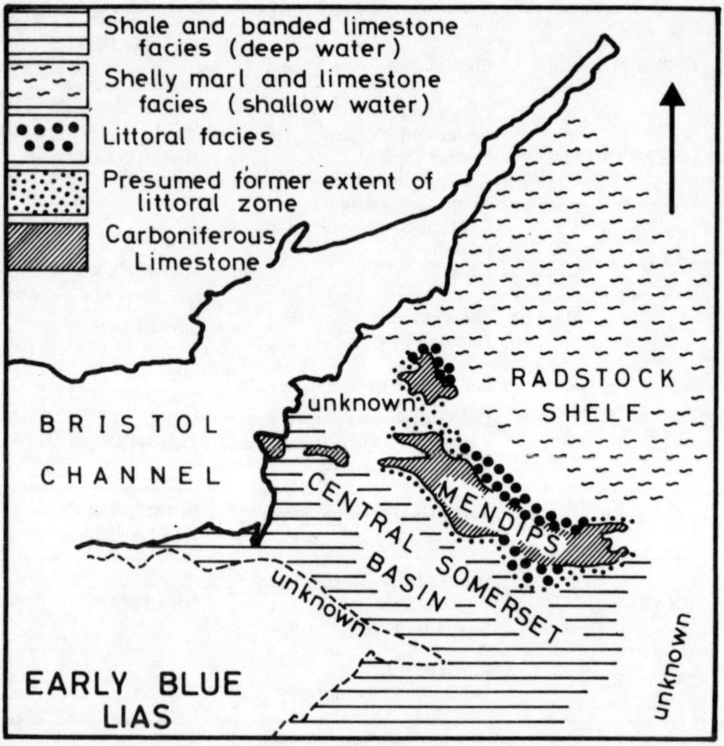

Fig. 2 Sketch map of the facies of the Lower Lias (After Kellaway & Welch, 1948; Crown copyright; reproduced by permission of the Controller, HMSO.)

Mendip are the rubbly Cornbrash limestones (loc. 11); by this time Mendip island no longer influenced deposition and it was submerged completely in later Jurassic times, though no later history is preserved until Pleistocene times.

Itinerary Leave Bristol going south on A 38. Two miles from the city centre along the road across Bedminster Down a stop can conveniently be made in the layby with public toilets. The view to the north from here extends across the Coal Measure basin of Ashton Vale to the Avon Gorge, cutting the Carboniferous Limestone peneplain which stands at around 100 m OD. The near horizontal Lower Lias of Bedminster Down stands at 70 m OD and to the SE Dundry Hill rises to 233 m with a capping of Inferior Oolite (Dundry Stone).

1 **Barrow Reservoirs** (ST 534 673) Proceed along the A 38 and at 3.6 km beyond the city boundary pull in at the Silver Star transport café on the right. Here is seen well exposed 3 m of normal facies of Blue Lias (Planorbis Zone) with typical alternating calcareous mudstones and clays.

2 **Brockley Combe Quarry** (ST 494.662) Continue south on A 38 for 2.8 km to Lulsgate Bottom; turn right (west) at the New Inn on minor road signposted Downside and Brockley Combe; proceed 2.2 km along this road to the entrance on the left of Downside House Farm (signposted North Somerset Welding Services). Permission to visit the site should be obtained from

Mrs. Cook (see p. 85); the quarry, which is to be seen in the field on the left of the entrance, is now being infilled but a small section may remain visible. About 4.5 m of massive porous detrital limestone is to be seen on the quarry face; it is sometimes known as Brockley Down Stone and was quarried as a freestone and used in Downside House and village. Bivalves recorded from the quarry include *Ceratomya, Gervillella, Liostrea, Modiolus* and *Parallelodon* (Donovan 1958a). The freestone is a littoral facies of the Lower Lias (Hettangian); it is less than 5 km from the previous stop where normal Lower Lias facies was seen. In early Liassic times Broadfield Down probably formed a shoal or bank in a very shallow sea, and on this bank accumulated the littoral facies with a conglomeratic base. A similar facies is to be seen at the Viaduct quarry, Downside, Shepton Mallet (see loc. 8 below). The duplication of the place-name Downside is a coincidence; both sites are recorded as having 'Downside Stone'; fortunately they are of the same age and lithology.

3 **Wurt Pit** (ST 5588 5391) Return to the A 38 at Lulsgate Bottom, continue S for 6 km and take minor road to left signposted Burrington; travel on B 3134 up Burrington Combe and continue as far as the Castle of Comfort Inn (ST 544 532). From here take road NE towards East Harptree. At the first bend as the road begins to drop down hill, a pit is visible in the field on the right, near the edge of the wood. Permission to visit the pit should be obtained from the owner (see p. 85) and the field entered by the gate near the corner of the wood.

The geomorphology of the pit is described in Excursion 6, loc. 10; it is a collapsed solution cavity presumably originating in the underlying Carboniferous Limestone. The pit is over 12 m deep; the top 3 m comprise massive chert with rich shelly fauna (see Donovan 1958a). The fossils include the ammonite *Psiloceras*: bivalves *Astarte, Cardinia, Lima, Liostrea, Modiolus, Pholadomya, Pinna* and *Pleuromya;* and the coral *Coelostylina.* These taxa are characteristic of the basal beds of the Blue Lias in the district and interest centres on the silicified aspect of the facies. The Geological Survey refer to the silicified Lower Lias and Inferior Oolite succession as the Harptree Beds. Rhaetic sands and silicified Keuper are also recorded in the district; they appear to be the result of metasomatic replacement (Green & Welch 1965).

4 **Emborough Quarries** (ST 623 505) On leaving Wurt Pit return to the Castle of Comfort Inn and turn south along B 3134 for 1 km to the Miner's Arms crossroads; bear east here along the B 3135 and continue on this road for about 5 km until it crosses the B 3139; turn NE along B 3139 for 3.5 km to Old Down crossroads. Some 200 m short of the crossroads the entrance to Emborough quarries is seen on the south side. Permission must be obtained in advance to enter quarries, see p. 85. Detailed geology of the Carboniferous Limestone will be found in Green & Welch (1965) and of the fissure infill in Robinson (1957).

Two quarries straddle the old railway line (see Fig. 3). The fissure survives as a prominent abutment in the SE corner of the eastern quarry; here the Hotwells Limestone (Zone D) dips towards the NE. The knoll is mostly infill; water-worn faces and swallet holes filled with red clay can still be seen. The lower part is well bedded red clays while the upper part is composed of a conglomerate with large boulders of Carboniferous Limestone set in red silty matrix. Bones occur in the conglomerate, usually totally dissociated; they mostly belong to a small gliding lizard *Kuehneosaurus.* Robinson (1957) interpreted the deposit as a collapsed cave; sediments in the lower part were brought in by an underground watercourse; later the thin limestone roof collapsed and streams and rainwash brought in silts with lizard remains. The age of the fauna and debris are pre-Rhaetic and probably late Triassic. Rhaetic sediments are recorded near the old quarry offices and in the subsoil on top of

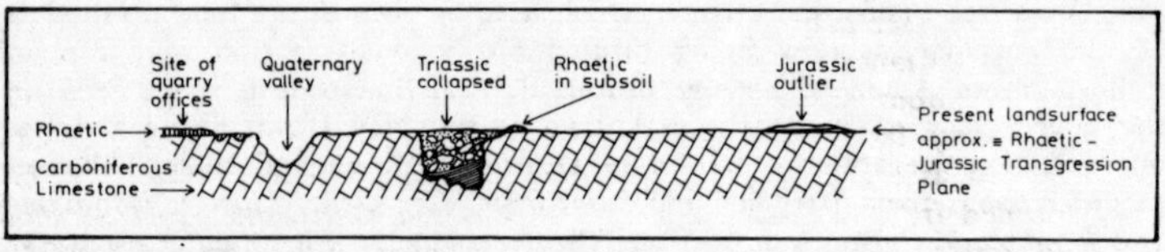

Fig. 3 Section across Emborough quarry (after Robinson 1957)

the limestone peneplain.

5 **Chilcompton Railway Cutting** (ST 632 512) Return to the B 3139 road and continue eastward for 1 km, stopping by the bridge where a descent to the old railway line can be made from the NW side of the bridge. The section seen immediately south of the bridge is now much overgrown and nowhere as good as that illustrated by Woodward (1876, Pl.IV, p. 79) – a lithograph based on a photograph taken by H. H. Winwood in 1873. Adjacent to the bridge on the SW face of the cutting the Rhaetic and Lias are folded into a small syncline (see Fig. 4) and are cut out against a fault which brings up the Dolomitic Conglomerate. The succession is given in Woodward (1876) and Reynolds (1911) recorded:

		m
Lower Lias	Clays and limestones	
Rhaetic	Langport Beds	1
	Cotham Beds	1.5
	Westbury Beds	2
Keuper	Tea Green Marls	
	Red Marls	
	FAULT	
	Dolomitic Conglomerate	

Nowadays only the Langport Beds (cream limestones) and the Dolomitic Conglomerate are exposed. The Dolomitic Conglomerate is rich in calcite and siliceous geodes. In the village of Chilcompton along Fry's Well (turning at Redan Inn) the Dolomitic Conglomerate is freshly exposed in a private garden by the railway bridge (ST 649 515).

6 **Gurney Slade** (ST 6235 4995) Return to Old Down and take A 37 south toward Shepton Mallet. After 1.3 km just as the road begins to drop down to the village of Gurney Slade a dyke is seen in a field on the right of the road. The site has been described by Moore (1867) and Robinson (1957). Already in Moore's time the Carboniferous Limestone had been quarried away and the vertical wall of fissure infill left standing (see Fig. 5). The dyke is orientated E–W, is some 4 m high and just over 1 m wide. The outer lining is composed of red stained often unbedded Triassic sediment, with an inner case of Liassic limestone of typical Downside Stone facies (see loc. 8 below) with essentially horizontal bedding. The latter facies has yielded the brachiopod *Calcirhynchia* and molluscs *Modiolus, Plagiostoma* and *Pleurotomaria;* the age is probably Upper Hettangian (Lower Lias). No vertebrates have been recorded. Numerous veins and pockets of hematite occur throughout the fissure. Robinson has interpreted the structure as a neptunian dyke; she envisaged an open joint in the Carboniferous Limestone when the Lias seas swept across the area and into which the sediment sagged.

7 **Windsor Hill** (ST 6145 4515) Continue south toward Shepton Mallet for about 5.5 km to the fork at Downside for Shepton Mallet town and the bypass (ST 624 448). Here take the narrow lane to the west (unsuitable for coaches) and follow the route shown in Fig. 6.

The site is described in Kühne (1956) and Robinson (1957). The disused

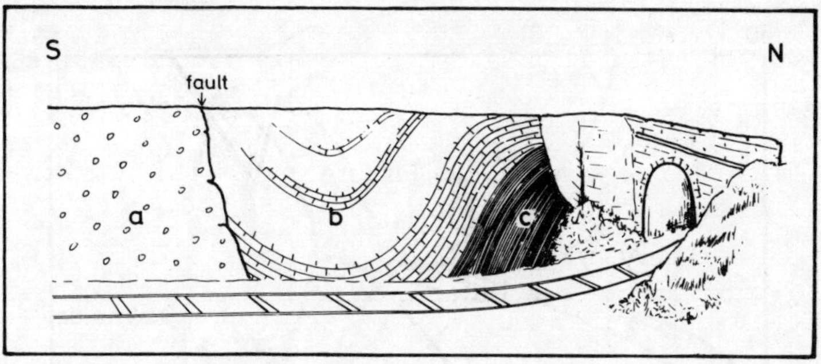

Fig. 4 Section in Chilcompton railway cutting a – Dolomitic Conglomerate; b – Lias; c – Rhaetic

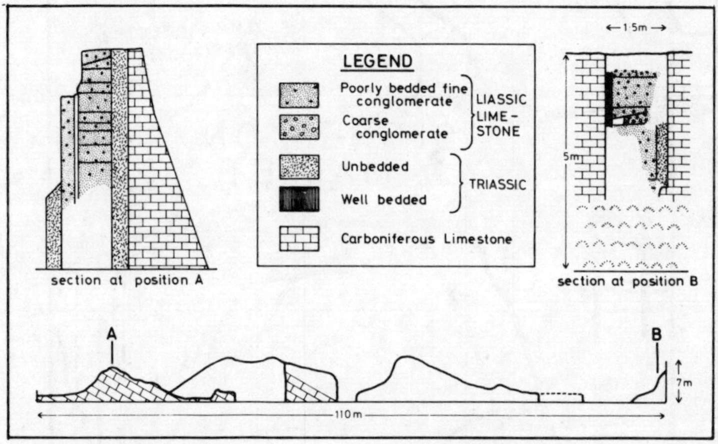

Fig. 5 The western part of the Gurney Slade fissure. Lower diagram: fissure seen from the south; Mesozoic rocks left blank (After Robinson 1957)

quarry is in Black Rock Limestone (Zone Z) which dips 35° to SSW. The essentially E–W fissure roughly follows the strike in the SW corner of the quarry where about 50 m of it are exposed at intervals along the quarry wall. The fissure is narrow, usually less than 1 m in width; the infill comprises pebbly, ferruginous cream-brown sandy limestones; where bedding is seen it generally dips at about 10° to the west. Most (90%) of the pebbles are Carboniferous chert and limestone. Some patches of the infill are very rich in bone; Kühne calculated up to 200 bones per cubic foot. All the bones are dissociated but fresh and belong to a genus of mammal-like reptile *Oligokyphus,* which was about the size of a rabbit and had similar dental specialisations – gnawing anterior teeth and grinding cheek teeth. Animals similar to it are recorded from Holwell (see loc. 12 below) and from South Wales; while not on the main line of mammalian evolution, it was close to the ancestral mammals. Other faunal elements recorded include fish remains (*Acrodus, Hybodus* and *Birgeria*) and invertebrates, some at least of which are remanié and which comprise ammonites, belemnites, gastropods, bivalves and brachiopods.

Kühne (1956) concluded that the age of the infill was late Lower Lias on

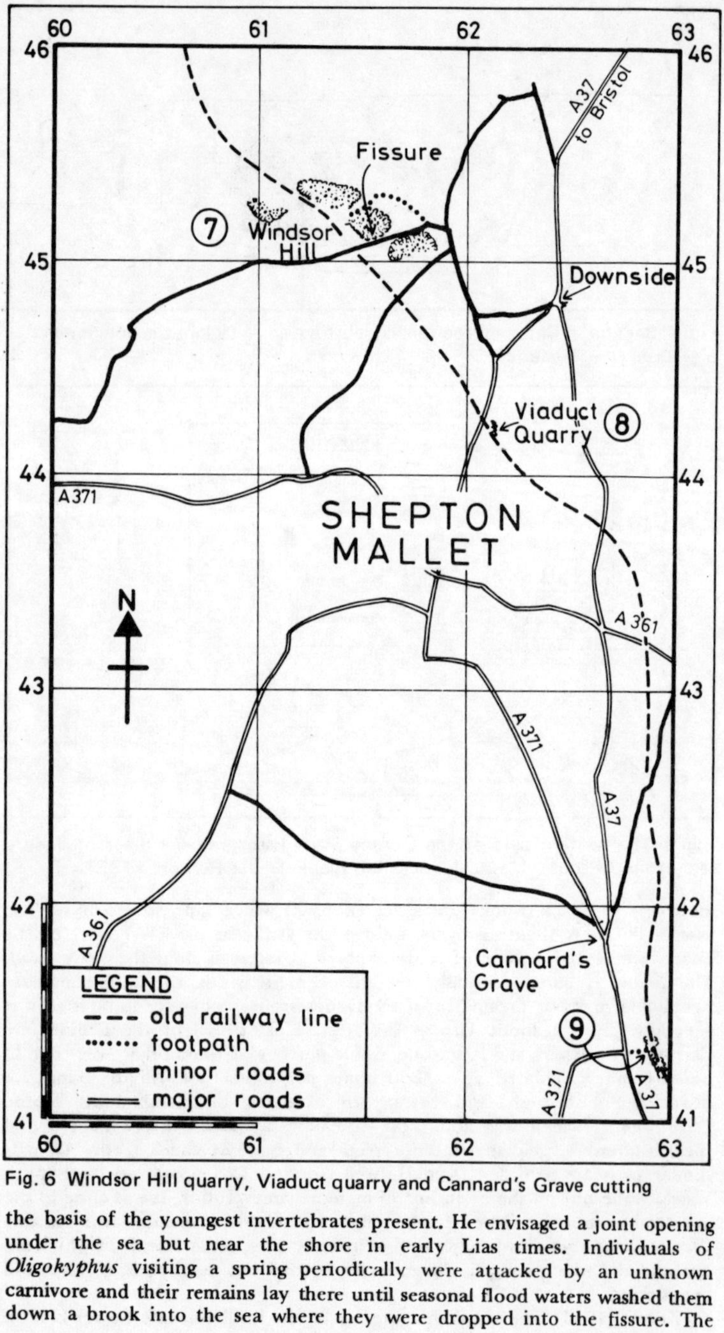

Fig. 6 Windsor Hill quarry, Viaduct quarry and Cannard's Grave cutting

the basis of the youngest invertebrates present. He envisaged a joint opening under the sea but near the shore in early Lias times. Individuals of *Oligokyphus* visiting a spring periodically were attacked by an unknown carnivore and their remains lay there until seasonal flood waters washed them down a brook into the sea where they were dropped into the fissure. The author has a different interpretation; he sees it as an infill from a swallet or grike in early Inferior Oolite times when the area was subaerial, with the inwashing of unconsolidated adjacent Lias sediments and contemporary remains of the reptiles.

8 Viaduct Quarry (ST 621 443) Return to A 37 and proceed south toward Shepton Mallet town centre. At a distance of 0.6 km from the Downside fork, the old railway viaduct is seen crossing the A 37; immediately north of the railway on the east side of the road is an old quarry in the Downside Stone, the littoral facies of the Hettangian. About 10 m is seen in the section, but nearby at Bowlish 16 m are recorded. The facies includes crystalline limestone, conglomerate with pebbles of Carboniferous Limestone and chert, and porcellaneous limestone. The ammonite *Psiloceras* has been recorded from the quarry, implying Hettangian age for at least part of the section (Donovan 1958a). A small working quarry in the Downside Stone is to be seen 1.5 km to the NE by the Fosse Way (ST 635 448). Compare these two sites with loc. 2 above.

9 Cannard's Grave (ST 629 414) Proceed through Shepton Mallet and continue south along A 371 (Cannard's Grave Road) to Cannard's Grave; stop 0.5 km south of here at the A 37/A 371 fork; a lane to the east of the road junction leads 200 m to a bridge over the old railway line. The cutting is best seen on the east bank, north of the bridge; however it is currently being infilled and may soon disappear. Donovan (1958b) recorded Lower Lias clay, Blue Lias limestones and shales, and Langport Beds. At present only the Blue Lias is visible and it has yielded *Psiloceras* (*Caloceras*), *Alsatites, Schlotheimia* and coroniceratids. The section displays typical Lias lithology in contrast to the littoral facies seen at the previous locality (8) 3 km to the north in Shepton Mallet. The normal succession at Cannard's Grave of 7.3 m of Blue Lias contrasts with thicknesses of 24 to 30 m for the same zonal sequence at Shepton Mallet. This thickness of pebbly bioclastic limestones of the Downside Stone facies is attributed to the banking up by offshore currents, where a condensed succession might otherwise be expected (Green & Welch 1965).

10 Doulting Railway Cutting (ST 647 425) and **Doulting Quarry** (ST 647 435) Return to A 37, turn north on the bypass road signposted Bristol; at traffic lights turn east along A 361 toward Frome; 1.2 km beyond traffic lights fork right on minor road for 1 km to junction at railway bridge. From here the railway cutting can be entered from the SE side of the bridge. Doulting quarry is 1 km north of the railway bridge, just beyond the A 361 road; permission to visit the quarry should be obtained from the Manager (see p. 85).

Richardson (1907) recorded the detailed succession in the railway cutting and the quarry; much of the lower part of the sequence in the cutting is now obscured. More recently Torrens (1969) and Parsons (1975) have recorded their interpretations, on which the section overleaf is based.

The Upper Lias is rarely seen on the Mendips; it is recorded around the Doulting area but not permanently exposed nowadays. There follows a non-sequence with as usual on the Mendips the whole of the Lower and Middle Inferior Oolite missing. The Doulting Conglomerate marks the frontal advance of the Upper Bajocian transgression; it is succeeded by the Ragstone and Doulting Stone, massive freestones with crinoidal content in the lower levels grading up into oolitic limestones. The Ragstone has some resemblance to the Downside Stone (see loc. 8). but the crinoids here may be contemporary and not derived from the Carboniferous. A well marked hardground level is easily seen in the quarry; it represents a sedimentary break during which time the surface was colonized by epi- and in-faunal communities. (The ecology of another Middle Jurassic hardground 25 km to

Zones				m
ZIGZAG	Lower Fuller's Earth	Knorri Clays	Brown-yellow clays with abundant small oysters (*Catinula*) near base	0.6
		Fullonicus Limestone	White argillaceous cementstones and brown marls: fossils include *Pholadomya* and *Procerites*	1.2
PARKINSONI	Upper Inferior Oolite	Anabacia Limestones	Rubbly oolitic brown-white limestones with small button coral '*Anabacia*' throughout: top surface bored and iron-stained	2.5
GARANTIANA		Doulting Stone and Ragstone	Massive wedgebedded bioclastic limestones, oolitic grading down into crinoidal: bored surface with hard ground characters; casts of trigoniids	13.5
SUBFURCATUM		Doulting Conglomerate	Grey crystalline limestone with pebbles: serpulids and borings abundant: plentiful terebratulids (*Sphaeroidothyris*)	0.4
			(Non-sequence)	
TOARCIAN	Upper Lias	Clays	Bluish, micaceous and arenaceous	0.6

Stratigraphical succession at Doulting

the NE at Bradford-on-Avon is described by Palmer & Fürsich (1974)). Doulting Stone has been quarried at least since the Middle Ages and was used in the building of Wells Cathedral, Glastonbury Abbey and all the older buildings of Doulting village. The Anabacia Limestones which succeed the Doulting Stone are rubbly and oolitic, and throughout can be found the small button coral 'Anabacia' (= *Chomatoseris porpites* (Wm. Smith)). Their top surface is bored and iron-stained with infillings of the overlying Fuller's Earth beds, comprising clays, marls and limestones. The combination of the railway cutting and the quarry enable the visitor to see much if not all of the details in the succession.

11 Merehead (ST 694 436) Return to the A 361 and continue eastward for 4.5 km to the Merehead quarry entrance; permission to examine the sections should be obtained from the quarry offices (see p. 85).

The first section (Fig. 7 (A)) is seen by the quarry entrance on the west bank of the railway line where it crosses the A 361 and on the north side of the road. Here a small fissure in the Carboniferous Limestone contains fish and brachiopods that indicate Sinemurian (Lower Lias) age; the section is very small and visitors are asked not to use hammers. Immediately east of the quarry entrance a bank alongside the disused road exposes well-bedded Anabacia Limestone which grades up into rubbly Fuller's Earth rock with abundant brachiopods near the top (B). A third section (C) is seen along the quarry railway line to the south (ST 684 433) where the Forest Marble is downfaulted against Fuller's Earth, and Cornbrash can be seen by the railway bridge (ST 698 430).

In the Doulting area (loc. 10) it was demonstrated how in Upper Bajocian (Upper Inferior Oolite) times the sea transgressed along the fringe of the Mendip hills. The effects of the Mendip fold were still marked in the succeeding Gt. Oolite times. The thinning out toward and across the Mendips is seen from three sequences (Arkell 1933) tabulated below.

	Wincanton	Batcombe	Frome	
Upper Fuller's Earth	73	41	6	thickness in metres
Fuller's Earth Rock	11	8	2	
Lower Fuller's Earth	36	6	2	
	120	55	10	

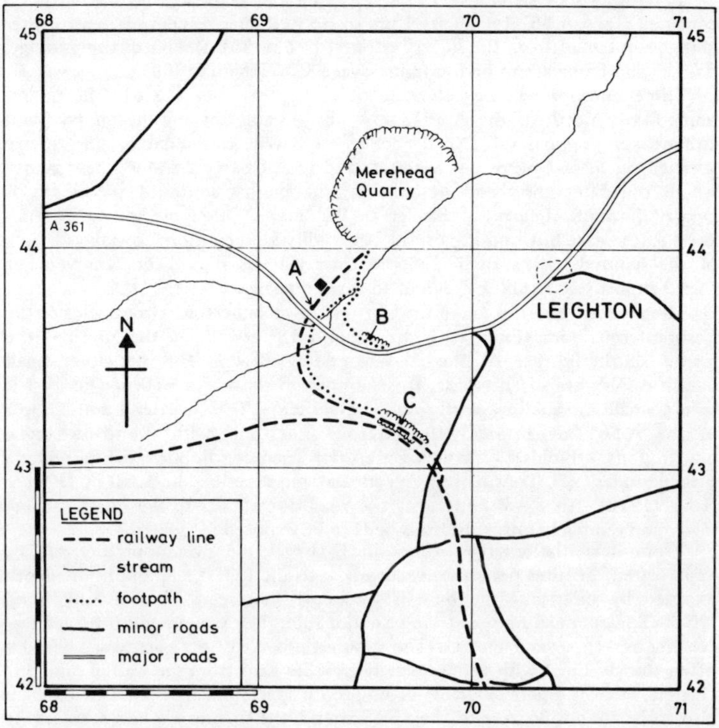

Fig. 7 Merehead quarry localities

Wincanton is 16 km south of the Mendip front and has a normal thickness of beds; Scale Hill near Batcombe is 5 km south of the front and a considerably reduced thickness; Bonneyleigh Hill near Frome is on the Mendip fold and has only a thin skin of sediments. The succeeding Forest Marble however is unaffected by the fold and thicknesses remain around 30 m at Merehead.

Around Merehead the Fuller's Earth Rock comprises rubbly limestones and clays; some of the limestone bands are rich in ornithellid brachiopods. The Fuller's Earth clays are not particularly rich in fossils; oysters are fairly common. The Forest Marble is one of the most characteristic facies in the Jurassic. The name was applied by William Smith to beds in the Forest of Wychwood in Oxfordshire; it is not a marble and does not represent a fossil forest although wood fragments are fairly common. The beds comprise clays and limestones; the limestones are shallow water current-bedded oolitic and sometimes sandy, blue when fresh, weathering to yellow-brown; they are detrital limestones made up very largely of comminuted shell fragments and when split along muddy interbeds yield abundant complete shells; oysters (*Liostrea*) are commonest, but there is a very wide spectrum of bivalves (e.g. *Camptonectes, Placunopsis*), brachiopods, bryozoa, crinoid ossicles, echinoid spines, sponges, fish scales and plant remains. The clays are bioturbated and commonly have traçe fossils. All the complete oyster shells in the interbeds are convex uppermost, while in the limestones the shells are fragmented; the indications are of a fairly high energy death assemblage in a marginal marine environment.

12 Holwell Quarries (ST 727 452) From the Merehead quarry entrance continue along A 361 for about 4 km to Holwell quarries; permission to enter must be obtained from the quarry offices, (see p. 85). Details of the geology, fissures and fauna are to be found in Savage & Waldman (1966).

Three quarries will be referred to in the text (see Fig. 8). The quarry immediately north of the A 361, now disused and housing the quarry plant and offices, is Quarry 1 (North quarry of Savage & Waldman); the quarry situated north of Quarry 1 is here referred to as Quarry 2 and is a new quarry which postdates the 1966 description; the quarry south of the A 361 is Quarry 3 (South quarry of Savage & Waldman). Other overgrown quarries exist to the east but nothing useful can readily be seen there nowadays. Much of the fissured rock seen in 1966 in Quarry 1 has since been removed but useful exposures can still be seen in all three quarries.

The Holwell quarries are in the Clifton Down Limestone Group (S_1) of the Carboniferous Limestone. Dips are around 12° to 25° to the south – the gentle southerly face of the E–W trending Beacon Hill pericline; small parasitic folds are often visible, for example on the north wall of Quarry 1 is seen a small and shallow south plunging anticline. The Quarries 1 and 2 north of the A 361 are in richly fossiliferous limestone, with abundant corals, brachiopods, crinoids and bryozoans. A rich seam of trilobites was found by a Bristol student (N. Trewin) some years ago and described in detail by Hahn & Hahn (1973). An E–W fault near the road downfaults to the north causing lower and relatively unfossiliferous beds to be exposed in Quarry 3.

From near the quarry offices in Quarry 1 the overall picture can be appreciated. Fissures penetrate vertically into the limestone, and in the depth exposed by quarrying the base is frequently not seen. They usually trend WNW–ESE, paralleling the strike of major fold; they vary in width from a few centimetres to several metres. The fissures, especially the narrower ones, are often thickly lined with calcite which grew inwards from the wall in radiating clusters, as beef (sparry calcite) or dog-tooth spar; sphalerite and hematite are also to be found. Because of their E–W trend, the fissures are best seen on the east and west faces of the quarry; several still survive in Quarry 1 though dust from the quarry plant makes them difficult to distinguish. The top of the Carboniferous Limestone is peneplained and where exposed usually displays remains of epifaunal (e.g. oysters) and infaunal (e.g. boring molluscs and worms) biota. Over this rock platform the early Mesozoic seas swept. On the north face of the quarry the top is seen to be capped by a few metres of rubbly Inferior Oolite; a N–S fault on this face displaces the Inferior Oolite, giving a thicker succession on the downthrown east side.

Moving to Quarry 2, the southern face is exclusively in Carboniferous Limestone, but an E–W fault through the quarry brings the unconformity within the quarry. It is usually possible to see the hardground in Carboniferous Limestone well exposed with mollusc and worm borings and oyster colonies. The Inferior Oolite comprises sandy oolites which increase in thickness as the ground rises to the north. The oolites are rich in fossils; valves of the lamellibranch *Entolium* are very common and casts of *Trigonia, Lopha, Pseudolimea* and *Pholadomya* are also to be found. Terebratulid and rhynchonellid brachiopods are not infrequent and there are abundant fragments of the zonal ammonite *Parkinsonia parkinsoni*. Fissures are usually to be seen in this quarry; those which do not penetrate the Inferior Oolite are certainly Mesozoic and those that penetrate to the ground surface are Pleistocene in age and usually have yellow-brown clayey infillings. No vertebrates have yet been reported from the quarry.

Turning now to Quarry 3 south of the A 361, it is immediately noticed that erosion has removed the whole of the Mesozoic cover. The most striking feature is a fan-shaped mass of Dolomitic Conglomerate which stands proud as

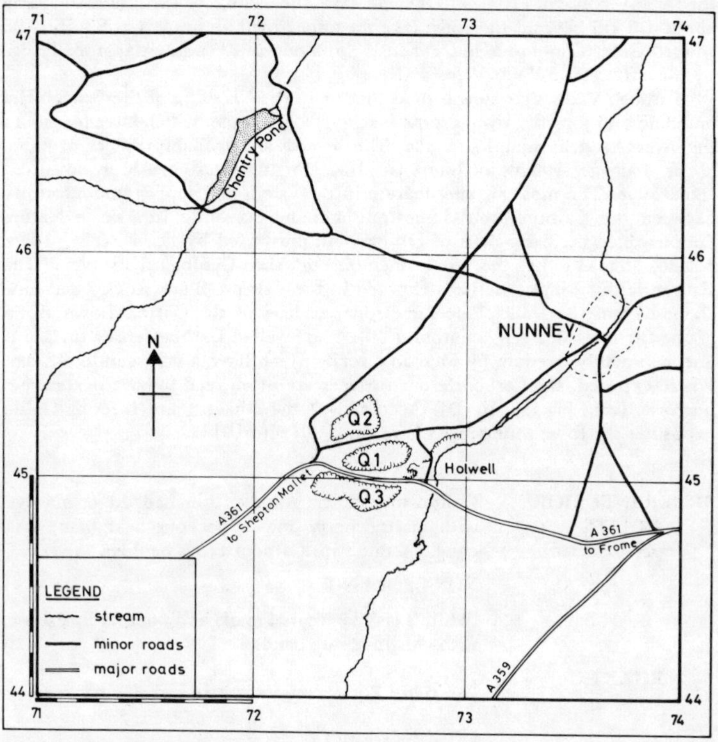

Fig. 8 Holwell quarries

a large promontory on the south face of the quarry. The geometry of the conglomerate mass as seen during quarrying suggested a source to the north; the boulders are mostly of Carboniferous origin and well rounded indicating probable transport down a wadi in the rainy season. The Dolomitic Conglomerate fan is seen on its west face to be cut by several E–W fissures, usually only 10–50 cms across and infilled with greyish clays containing Rhaetian fish teeth. The Dolomitic Conglomerate must have been lithified between its deposition in Triassic times and the formation of the fissures; the infillings are not earlier than Rhaetic and may be as late as early Inferior Oolite. On the north wall of this quarry a small cave system is seen to be filled with red bedded marls, which have not yielded any vertebrate remains. In addition to these Mesozoic infillings, the southern face of the quarry usually displays some Pleistocene fissure infillings similar to those in Quarry 2. It was from Quarry 3, near the entrance, that fissures now quarried away yielded to Mr. Charles Moore of Bath in the 1850's and 1860's not only large quantities of Rhaetian fish teeth (over 70,000 specimens of *Acrodus* were recovered by him) but also no less than 15 mammalian teeth which he named *Microlestes* (Moore 1867) and which are now known under the name of *Haramiya*. Kühne (1946) washed a further two tons of fissure infill and found another 20 mammalian teeth. These teeth rank amongst the earliest known mammalian remains. Savage in the 1960's found in a fissure in Quarry 1 the jaw of a mammal-like reptile species close to the Windsor Hill *Oligokyphus*.

13 **Vallis Vale** (ST 755 490) Leaving Holwell quarries, continue east on

the A 361 toward Frome. In Frome take the A 362 toward Radstock and about 1.5 km beyond the town take the minor road on the left at ST 765 496 toward Great Elm and Mells. Park 0.5 km along this road at Hapsford bridge by the entrance to Vallis Vale.

Through Vallis Vale flow brooks that drain the NE slope of the Beacon Hill pericline and join the river Frome just north of Frome, which later feeds into the river Avon at Limpley Stoke. The brooks are probably relicts of much larger drainage systems of late Glacial age for the sides have been down cut some 50 m. The most striking feature in the vale is the superb unconformity between the Carboniferous Limestone and the overlying Jurassic, a feature famous since the early days of geology and illustrated by de la Beche (1846, p. 288, fig. 41) in the first memoir of the Geological Survey. The Carboniferous Limestone beds exposed in the Vale are Black Rock Limestone (Zone Z) and the Vallis Vale Limestone, a facies of the Clifton Down group (Zone C); beds dip NW at around 30°. The levelled Carboniferous surface is unconformably overlain by Mesozoic rocks which have a very gentle NE dip. Where exposed, the Carboniferous surfaces are often seen to have oysters and burrows (e.g. Fig. 9, site **D**). Accounts of the Rhaetic and Inferior Oolite exposures are to be found in Richardson (1907 and 1911).

	Section in Vallis Vale
UPPER INFERIOR OOLITE (Upper Bajocian)	Oolitic limestones; rubbly, thin bedded or massive, with marls; cherty and arenaceous near base; basal conglomerate with Carboniferous pebbles.
	NON-SEQUENCE
RHAETIC 0–4 m	'White Lias'; laminated marls and nodular limestones with conglomerate bands.
	'Westbury Beds'; conglomerates and clay partings
	UNCONFORMITY
CARBONIFEROUS	Vallis Vale Limestone

Stop A has the most complete succession with most of the Rhaetic present, though much of the upper half is overgrown and not readily accessible. At stop **B** about 150 m to the south along the brook the Rhaetic is only about 1 m thick. At both stops the Rhaetic rests directly on the Carboniferous Limestone and contains conglomerates with blackish clay partings. The conglomerates contain pebbles of Carboniferous origin which have been bored by molluscs during their exposure on the peneplain and before their incorporation in the conglomerate. The clays are sometimes rich in plant remains (*Lycopodites*) and have also yielded the characteristic Rhaetic bivalve *Rhaetavicula contorta,* together with the small crustacean *Euestheria minuta* and fish remains. The Moore collection in Bath contains a small dinosaur vertebra from these beds. A further 150 m southward along the old quarry faces at stop **C** the Inferior Oolite is seen resting directly on the Carboniferous Limestone; this situation maintains for much of the rest of the vale and the best exposure is at stop **D**. Here the surface of the Carboniferous is seen well exposed with borings made by the mollusc *Lithophaga* and colonized by oysters. The Inferior Oolite above the unconformity comprises oolitic limestones; rubbly, thin bedded or massive, with marly partings. The facies is very similar to the Doulting Stone facies (see loc. **10** p. 93) and has yielded a wide range of fossils including the bivalves *Ctenostreon, Pholadomya* and *Lima;* brachiopod *Acanthothiris;* echinoids *Cidaris* and *Acrosalenia.* Beyond this point further exposures of the unconformity may be seen by following

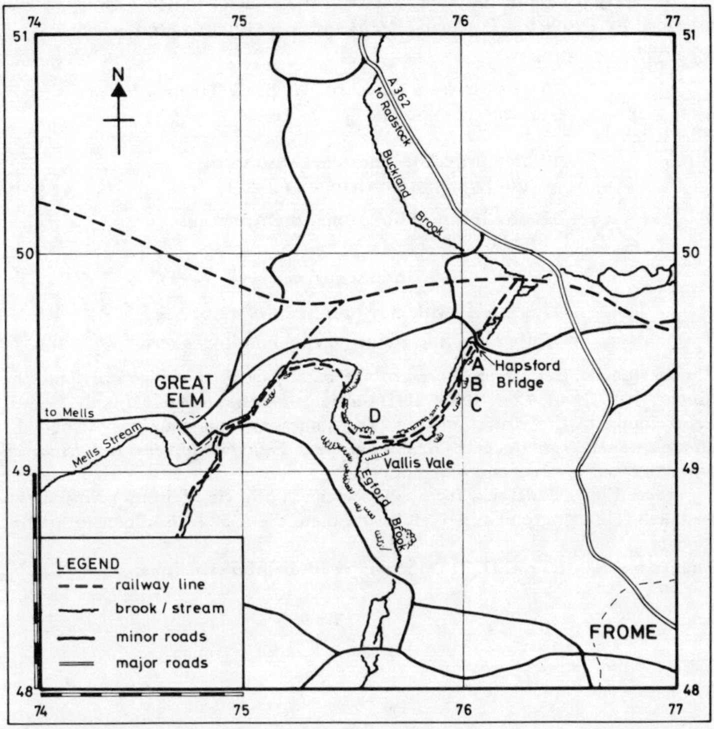

Fig. 9 Vallis Vale localities

the Great Elm branch of the brook to the west and along the Egford brook to the south.

On leaving Vallis Vale by the Hapsford bridge entrance, the traveller is recommended to retrace his journey eastward for 0.4 km where on reaching the A 362 turn north (left) towards Radstock.

14 **Huish Colliery Quarry** (ST 696 543) About 1 km before reaching the centre of Radstock at the Fromeway Inn turn left along Mells Lane. After about 300 m where the houses end, leave transportation and take a narrow path on right between two houses, 'Newlands' and 'Bellevue'. The path is an old tramway incline; descending a grassy slope, the quarry is found on the left as the path cuts through shrubbery.

The Lias of the Radstock basin is characterised by its thinness and rich fossil assemblage. The Lias of Gloucestershire thins southward to a minimum along the north fringe of the Mendip Hills; in a similar way the Lias of Dorset thins northward to a minimum along the south fringe of the Mendips. Formerly the details of the succession could be followed in many small quarries around Radstock, but all these have now ceased working and most are very overgrown or otherwise inaccessible. The penecontemporaneous movement of the Mendip axial fold during early Jurassic times has resulted in numerous non-sequences in the succession; these are often marked by erosion planes, by bands of phosphatized nodules and beds of remanié fossils. Sometimes no marked stratigraphical break is detectable and it is only palaeontologically that the non-sequence is recognisable.

The section here in the shelly facies of the Radstock Shelf Lower Lias was recorded by Tutcher & Trueman (1925) and is as currently seen:

STAGE		m
Pliensbachian	Clays ('Striatum Clays' of Tutcher & Trueman) formerly exposed	
	Rubbly, ironshot limestones with fossils (the Jamesoni Limestone of T & T) seen	1.6
	Massive ironshot limestone (the Armatum Bed of T & T)	0.4
	------------- non-sequence ------------	
Hettangian	Limestones with clay partings, *Psiloceras*	1.3
	White Lias: fine grained cream limestones seen	0.6

The Hettangian limestones represent the basal part of the stage only, and the Sinemurian, about 42 m thick at Dundry, is absent, remanié fossils from it being found in the Armatum Bed. The Jamesoni Limestone has yielded the ammonites *Platypleuroceras, Radstockiceras, Trugophylloceras* and *Uptonia,* and numerous bivalves and brachiopods.

Proceed into Radstock, from whence the A 362 via Midsomer Norton can be taken to Farrington Gurney and from there the A 37 leads direct to Bristol.

Acknowledgments

Thanks are due to Prof. D. T. Donovan for his helpful criticisms.

10

Coastal exposures near Blue Anchor, Watchet and St. Audrie's Bay, North Somerset

D. Hamilton and A. Whittaker

Objectives

The purpose of the excursion is to examine the fine coastal outcrops which expose the transition from evaporite-bearing Keuper and Tea Green Marls, through a full development of fossiliferous Rhaetic beds to Liassic marine shales and limestones. Tightly folded and faulted structures, with gypsum veins, are well exposed in the cliffs and wave cut platform. Deformed Pleistocene gravels occur in Doniford Bay.

Special Features

Low tides are essential for this itinerary. Since high tides reach the base of the coastal cliffs, visitors must take care not to be caught by the incoming tide. Excursions should start at the beginning of an ebb tide. Access to St. Audrie's Bay is by a private road to a car park and a toll is charged.

Maps

Ordnance Survey	One-Inch Sheet 164 (Minehead)
	1:50 000 Sheet 181 (Minehead & Brendon Hills)
	1:25 000 Sheets ST 04, ST 14
Geological Survey	One-Inch Sheets 294 (Dulverton), 295 (Taunton)

Main References

Bradshaw & Hamilton (1967); Richardson (1911).

Outline Geology

The north Somerset coast between Blue Anchor and St. Audrie's Bay exposes the following succession.

PLEISTOCENE	Sands and gravels.
JURASSIC	Lower Lias, shales and limestones.
TRIASSIC	Rhaetic, shales, limestones and marls.
	Grey Marl, mudstones, siltstones and gypsum horizons.
	Tea Green Marl, mudstones and siltstones.
	Red Keuper Marl, mudstones and siltstones.

The area lies on the southern shore of the Bristol Channel and to the north of the Brendon and Quantock Hills, which are composed of Devonian rocks. The coastal sections expose strata about the Triassic-Jurassic junction and illustrate the transition from a continental to a marine environment. Examples of superficial mass movement, such as landslips, are visible in some of the cliffs, and the morphology of the coast can be related to the solid geology. Many localities provide a good hunting ground for fossils and for the examination of faults, folds, joints, veins, sedimentary structures (ripples, dessication cracks, geodes etc) and cryoturbation structures in the Pleistocene gravels.

The Keuper sediments, mainly red calcareous and dolomitic siltstones, are interpreted as the water-laid deposits of an extensive Triassic plain, lying close to sea-level. The succeeding Tea Green Marls and Grey Marls are interbedded with abundant gypsum. These sediments were laid down either in hypersaline playa lakes or more probably on extensive tidal flats, under conditions of high rates of evaporation, giving rise to hypersaline groundwater conditions similar to the sabkahs of the Persian Gulf (Stevenson 1970). In contrast with the Persian Gulf sediments, those at Blue Anchor have a high detrital content. They thus indicate a sedimentary environment close to sea-level. The occurrence of marine influences in the higher Grey Marl sequence is also suggested by the presence of bivalves (Dawkins 1964) in the Sully Beds. It was also from the Sully Beds of this area that Dawkins recorded the earliest known British mammal, *Hypsiprymnopsis rhaeticus.*

The Rhaetic formation of the Triassic has ben subdivided into the following units by Richardson (1911).

			m
RHAETIC	Upper Rhaetic	Watchet Beds	1.5
		Langport Beds	1.0
		Cotham Beds	2.7
	Lower Rhaetic	Westbury Beds	13.8
		Sully Beds	4.0

Lithologically, and for the purposes of mapping, the Sully Beds are best classified with the Grey Marl sequence, although the occurrence in them of Rhaetic fossils led Richardson to group them with the Rhaetic. The Westbury Beds are predominantly dark grey or black shales but also contain current-concentrated shell beds and rippled sandstones with vertebrate remains and pebbles. These cyclic deposits (see Hamilton 1962) are probably tidal or sub-tidal. The shales and mudstones contain thin-shelled bivalves, including *Chlamys valoniensis* (Defrance) and *Rhaetavicula* (*Pteria*) *contorta* (Portlock). The Upper Rhaetic Cotham, Langport and Watchet Beds are thin and comprise pale grey limestones and grey or greenish grey mudstones. They are thought to represent deposits formed in a lagoonal environment. The exact position of the Triassic-Jurassic boundary in this part of the sequence has given rise to controversy.

Overlying the Upper Rhaetic beds is the alternating shale and limestone sequence of the Blue Lias. These marine strata contain abundant bivalves in their lower part and ammonites in the higher parts. Saurians are also recorded from the Blue Lias of this area. In the past, there has been much discussion as to whether the shale-limestone rhythm is a primary feature of the deposition or whether it is due to post-depositional segregation of calcium carbonate (Hallam 1964). Probably both processes have played a part in the formation of Blue Lias limestones. Fossils are generally not flattened in the limestones, suggesting a primary origin, but many limestone beds show irregular accretionary surfaces.

Though the Palaeozoic Brendon and Quantock Hills now rise to the south of the area, no coarse littoral deposits occur in the Mesozoic sequence; however, there is some evidence that the Lias thins in the proximity of the Palaeozoic rocks (see Whittaker 1973).

Important faults and folds trend ESE, although some structures are aligned NW (see Fig. 1). The ESE trend parallels structural directions in the Palaeozoic rocks to the south (see Webby 1965) and in submerged Mesozoic rocks

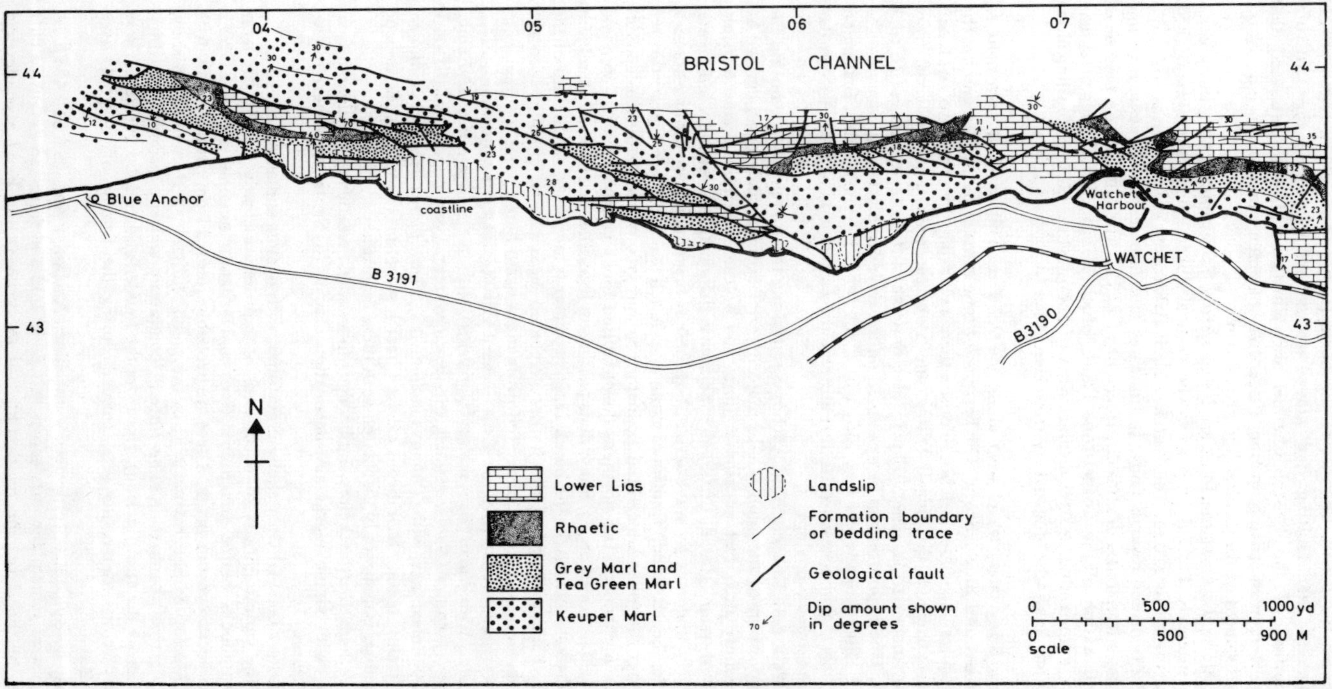

Fig.1 Simplified geological map of the intertidal area between Blue Anchor and Watchet. (Published by permission of the Director, IGS)

beneath the Bristol Channel to the north (Lloyd *et al.* 1973). Owen (1971) suggested that faulting had played an important part in the structural evolution of the Bristol Channel and Whittaker (1973) postulated that post-Hercynian rifting in the Palaeozoic basement may have led to the development of the Central Somerset Basin.

Along the coast the folds are long, narrow, ESE-trending structures which are intensely faulted in places. The vertical displacement along major ESE faults is up to 220 m. The Watchet Fault (Whittaker 1972a) trends NW and is a transcurrent reverse fault. It truncates and displaces E–W structures and shows a horizontal dextral displacement of 275 m. The fault lies on the line of continuation of the Cothelstone Fault, which Webby (1965) suggested may have a horizontal dextral displacement of about 5 km. Clearly, faulting along this line has had a long history from Hercynian to Tertiary times.

Itinerary

From Bridgwater, travel west along the A 39 to Williton then turn right to Watchet and Blue Anchor. The best use of time is made by first examining the exposures at Blue Anchor Point, then returning by road to Watchet. Alternatively, if the tide is falling, the party can walk along the coast from Blue Anchor to Watchet (4 km). Note that the cliff faces are subject to rock falls. Fine exposures can be examined on the wave-cut platform.

1 **Blue Anchor** (ST 034 435) Where the B 3191 from Watchet first meets the coast, gain access to the foreshore at the eastern end of Blue Anchor Bay by the slipway or steps.

Walk east towards Blue Anchor Point along the shingle beach, noting the composition of the pebbles. The majority are Devonian sandstones, derived by long-shore drift but occasional flints and volcanic pebbles occur, possibly derived from glacial drift. The adjacent cliffs expose Keuper sediments, actively undercut by wave action, but these soft sediments are rapidly abraded in the beach zone. Continue along the beach to the place where the red Keuper siltstones are faulted against green and grey marls. An extensive fault plane is exposed. The northern down-faulted block consists of Tea Green and Grey Marls overlain by Sully Beds passing up into Rhaetic strata (a prominent sandstone band marks the base). Liassic sediments are present at the top of the cliff. Fallen blocks of the harder cemented beds can be examined on the beach. Good specimens of the following often occur here:

Westbury Beds: (i) Bone Beds – hard gritty calcareous sandstones, often rippled, with small fish scales (*Gyrolepis*), fish teeth (*Acrodus, Hybodus, Birgeria*), spines (*Nemacanthus*), plesiosaur vertebrae and coprolites. (ii) Dark shelly limestones with *Chlamys valoniensis, Pleurophorus elongatus Moore,* fish scales, teeth, etc. Fibrous calcite ('beef'), sometimes showing cone-in-cone structure, is commonly associated with these limestones.

Cotham Beds: (i) Grey argillaceous limestones, weathering to light brown grey, with ripple marks and dessication cracks. (ii) Septarian concretionary structures.

Lower Lias: (i) Grey shelly limestones, with oysters and ammonites.

The various modes of occurrence of gypsum are clearly seen in the large fallen blocks beside the point. Beds and separate nodular masses of pink translucent gypsum occur. These are probably formed from primary anhydrite (cf. Mossop & Shearman 1973) in the green and grey siltstones, which show wavy (?tidal) lamination. The fibrous, commonly white, gypsum veins, which cut across the stratification, arise from the bedded and nodular gypsum. These veins are in intersecting sets, formed during the folding of the rocks (see Fig. 2A).

A conspicuous fold occurs at Blue Anchor Point (ST 040 437). A prominent joint plane cuts across this E–W fold and forms an extensive cliff face. In this face, the relationships of the veins to the E–W asymmetrical

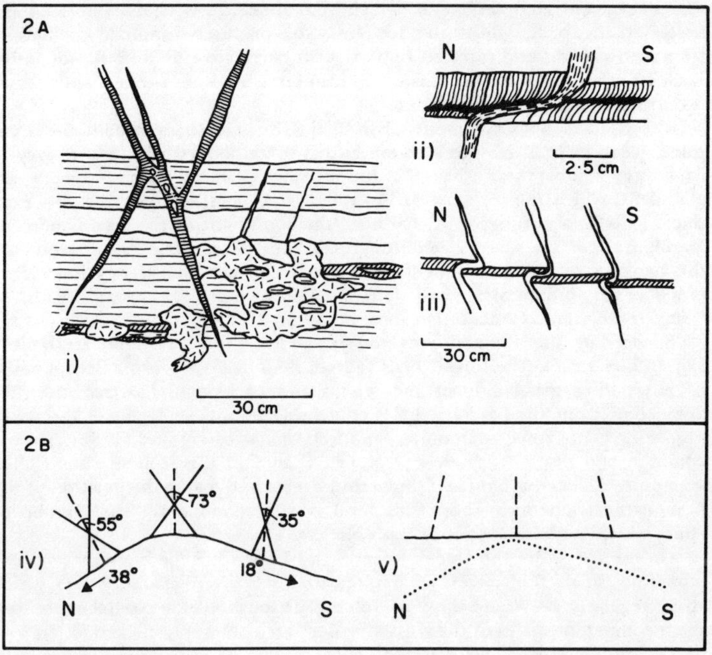

Fig. 2A Relationships of the various forms of gypsum at Blue Anchor Point.
(i) Nodular alabaster in the laminated beds, with white veins parallel and cross-cutting pink veins.
(ii) and (iii) Horizontal white veins cut by thin pink veins and both deformed by subsequent movements.
Fig. 2B Relation of conjugate gypsum veins to the anticline.
(iv) As seen at present. The bisector of the acute angle (direction of maximum compression) is shown by pecked lines.
(v) Orientation of the bisectors if the beds are returned to the horizontal. (After Bradshaw & Hamilton, 1967)

anticline (see Fig. 2B) can be clearly seen (Bradshaw & Hamilton 1967). The stratigraphic succession described in detail by Richardson (1911) can be examined in the wave-cut platform adjacent to Blue Anchor Point. Here, beds of the northern limb of the anticline dip towards the Bristol Channel. Several WNW trending faults cross this succession. The uppermost part of the Tea Green Marl and Grey Marl sequence occurs in the lowest 24 m of the cliffs. This sequence can be examined in detail in the wave-cut platform.

Some 4 m of Sully Beds follow, these being interbedded lighter and darker grey marls. Some of these have been disturbed by boring organisms, which are indicative of very shallow water or littoral conditions. Fossil fish scales and coprolites were reported by Richardson (1911).

The base of the Westbury Beds is marked by a 15 cm thick hard sandstone with vertebrate remains and fragments of the underlying Sully Beds. This is the lowest of three 'Bone Beds' occuring in the 13.8 m of Westbury Beds here. The others occur at 1.5 m and 6 m above the base. The other hard rock ribs are shelly limestones (biosparites) with abundant *Chlamys valoniensis, Pleurophorus elongatus, Rhaetavicula contorta* and fish remains. These are current-concentrated sediments which sometimes show channel fills on the bases of the beds, whilst the tops may be profusely rippled. Oscillation, truncated crested and terraced ripples occur, suggestive of tidal flat deposits. Each of these coarser beds passes up into laminated mudstones, with fossils occurring on some bedding planes.

The Cotham Beds are about 1.5 m thick. The limestones, usually 6–12 cm thick, weather to a light grey colour and are interbedded with brown grey to dark grey mudstones. Near the top of the Cotham Beds ostracods are abundant, whilst the topmost bed is a dark grey micritic limestone, 9–12 cm thick. This is, according to Richardson, the equivalent of the algal Landscape Marble horizon elsewhere (Hamilton 1961). The succeeding 3 m of strata are the Langport and Watchet Beds; they consist of thin bluish grey limestones (argillaceous biomicrites) and laminated brownish grey marls. *Liostrea hisingeri* (Nilsson) is common at some horizons.

Succeeding the Rhaetic beds and exposed on the foreshore, are faulted Lower Lias strata. The lowest beds (*Ostrea* Beds or *Preplanorbis* Beds) consist of alternating massive limestones and shales containing *Liostrea hisingeri*. Some 6 m from the base are shales containing smooth and plicate species of *Psiloceras*. The zonal ammonite *Psiloceras planorbis* (J. de C. Sowerby) is common and displays iridescent colours. Higher in the sequence the ribbed ammonite *Caloceras johnstoni* (Sowerby) occurs and marks the position of the *C. johnstoni* Subzone. Above this level is a predominantly shaly group of strata which contains species of *Schlotheimia*.

On the foreshore the Upper Rhaetic and Lower Lias beds are thrown into numerous small folds; they are faulted against Lower Rhaetic strata which strike regularly E–W and show no folding. Throughout the coastal exposures, the incompetent shales of the Westbury Beds are commonly strike-faulted.

On the NE side of the Watchet Fault (see Fig. 1), as exposed in the cliff, Rhaetic and Lower Lias strata are inverted. The beds dip towards the SW and yet young towards the NE. Red and green Keuper marls are in contact with Westbury Beds Shales, in places forming a fault gouge, and the main fault plane dips towards the SW at 55°. Diagonal slickensides are visible in some subsidiary fractures in the Keuper Marl very close to and parallel with the main fault. These indicate some horizontal component in the fault movements. The best illustration of the transcurrent nature of the fault is found on the foreshore at low tide. Here, the major E–W trending structures are truncated by the Watchet Fault and display dextral drag. The Watchet Fault, as exposed in the cliff, shows a vertical displacement of about 55 m, but this increases seaward. The horizontal displacement is demonstrated by matching outcrops across the fault plane and can be shown to be about 275 m.

In the cliff between the Watchet Fault and Watchet Harbour Lower Lias limestones and shales are faulted against red gypsiferous Keuper Marl. The fault parallels the cliff but is poorly exposed because it is covered by a clay sludge derived from the Lias. Gypsum in the Keuper Marl hereabouts occurs as nodular masses parallel with the bedding, and as veins. The type locality of the ammonite *Psiloceras planorbis* is to the north west of Watchet.

2 **Watchet** (ST 074 434) Gain access to the foreshore by the steps and path beside the railway line adjacent to the harbour. Walk eastwards along the strike of the Keuper Marls and examine the lithology and variegated red and green colours in the cliffs. About 600 m along the shore circular green spots, with a black carbonaceous core, occur in the red siltstones. Several small normal and reversed faults can be seen in these cliffs showing slickensides and cemented fault breccias. Sea caves are often eroded along these fault lines. In the variegated marls of this vicinity, a precursor of the Grey Marl-type lithologies is seen.

Before reaching the point, note the pronounced change in strike as the nose of the easterly plunging anticline is reached. A section across this structure is seen around the point, with an intersecting fault pattern. Trace out the structures immediately to the east of the point before climbing to the top to obtain a better view of the wave cut platform. En route note the change to light green colours of the Tea Green and Grey Marl on the south of the anticline adjacent to the concrete steps.

Return to the beach and examine the dark mudstones in the cliffs to the south. These dip to the north and yield abundant *Gryphaea*. The whole of the Rhaetic succession and much of the Blue Lias has been faulted out by the E–W Helwell Bay fault. About 100 m to the south, in the cliff, the mudstones and shales contain many, slightly crushed, specimens of *Arnioceras*. Near the wooden steps to the chalets, and on the foreshore, are limestones and shales which contain large, thick-shelled and coarsely-ribbed arietitid ammonites. Careful examination of the mudstones in this area will reveal many horizons, usually at the position of a lithological change, which contain trace fossils similar to *Chrondrites*. These are manifested as small 'spots' of different lithology (usually the overlying sediment type) in the particular stratum. Observations on trace fossils similar to these have been used as evidence to suggest that the limestone-shale rhythm is a primary feature of the sedimention. The Lower Lias beds in Doniford Bay are probably referable to the *Arnioceras semicostatum* Zone, and the major fault which crosses the bay downthrows about 213 m to the south.

Gravels of Pleistocene age overlying Liassic strata, are exposed in the top of the cliffs of Doniford Bay. The gravels are 3 m thick and have a sandy clay matrix. The pebbles and boulders show a series of sharp crested 'folds' or cryoturbations, formed by periglacial action. This gravel deposit is part of a more extensive spread which floors the valley between the Brendon and Quantock Hills. Only the lowest of the three divisions of these gravels is exposed in the shoreline cliffs. Palaeoliths, often highly abraded, of middle Acheulian age have been collected from these lowest gravels (Wedlake & Wedlake 1963). Gilbertson & Mottershead (1974) suggest that the gravels represent older Pleistocene deposits reworked by Devensian solifluction processes.

3 **St. Audrie's Bay** (ST 110 431) From Watchet proceed eastwards on the Doniford Road (unclassified) to Rydon Farm (2.5 km). At Rydon (ST 100 425) turn left down the private road which leads to Home Farm and the Holiday Camp. There is a small car park at Home Farm and a charge is made for parking and for each person visiting the beach. Coach parties may wish to rejoin their transport near Doniford where there is a caravan and car park conveniently situated near the cliff top.

From the Home Farm car park follow the signs to the beach down a zig-zag track. Upon reaching the beach, first proceed eastwards to the point near the waterfall to enjoy a view of the main part of St. Audrie's Bay. From here the headland of Blue Ben is visible to the east and Lias measures can be seen faulted against red Keuper Marl. The roughly E–W trending fault can be traced across the foreshore and for a considerable distance inland. It is a normal fault bringing shales and limestones of the *Arietites bucklandi* Zone into contact with red marls; the downthrow to the north is about 167 m. Still looking eastwards, a tussocky knoll, about 100 m inland from the cliff, near Perry Gully, is an inlier of Devonian Hangman Grit surrounded by Keuper Marl. The eastern end of St. Audrie's Bay foreshore displays small synclines in the Keuper Marl beds. However, from the place at which the observer is now standing, the beds strike regularly out to sea and the dip is fairly uniform at about 13° to the WSW. Hereabouts the reefs are formed by the hard, green and greenish grey siltstone bands, which are interbedded with the red marls near the top of the Keuper Marl sequence (Whittaker 1972b).

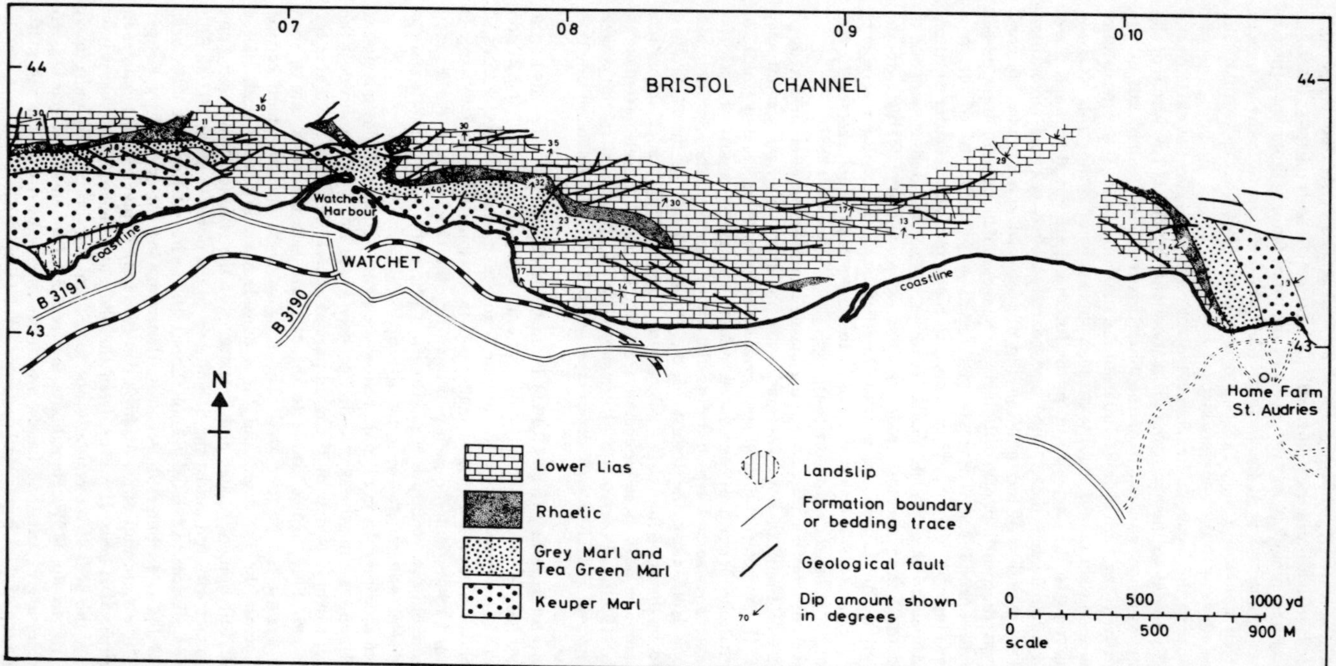

Fig.3 Simplified geological map of the intertidal area between Watchet and St. Audrie's Bay (Published by permission of the Director, IGS)

Proceeding westwards now, back towards The Slip and up the succession, it will be seen that the green bands become an increasingly important element in the stratigraphy. Just west of The Slip is the highest, prominent, red marl band which marks the junction between the red Keuper Marl and the Tea Green Marl. Hereabouts are also seen small high-angle reverse and normal faults. The Tea Green Marl (about 5.2 m thick) consists of green blocky and silty mudstones, shales and siltstones. The base of the overlying Grey Marl sequence (25.5 m thick) is marked by the lowest dark grey marl band: the Grey Marl beds comprise thin alternations of grey and greenish grey siltstones and silty mudstones, commonly with intercalations of dark grey marl which are prominent near the top of this division. As at the other localities, the top few metres (Sully Beds) of the Grey Marl contain Rhaetic fossils. Nodular gypsum horizons are poorly developed or absent at St. Audrie's Bay although small veins are present in places in the Grey Marls.

Succeeding the Grey Marls are Rhaetic beds. They display characters similar to those at the localities already mentioned above except that they are a little thinner. The soft Westbury Beds shales are easily eroded and this has led to local slipping at the western end of St. Audrie's Bay. Large masses of Blue Lias have been lowered to beach level and at the present time obscure parts of the Rhaetic sequence.

An interesting feature of the Cotham Beds is a massive bed near the base which shows contortions and convolutions of laminae and bedding thought to be due to penecontemporaneous slumping. The top surface of the same bed at St. Audrie's displays polygonal shrinkage cracks which are infilled, with material of different lithology, from a bed above.

Succeeding the Rhaetic and forming the western promontory at St. Audrie's Bay are the Blue Lias beds. The lowest division comprises alternating massive limestones and shales and corresponds approximately with the *Psiloceras planorbis* Zone. Near the base of the Blue Lias and at the foot of the cliff at the western end of the bay occurs the Bottom Lias Bed. This bed, which is involved in a little thrusting hereabouts, is a massive, fossiliferous limestone that can be recognized, along the strike, for considerable distances. Ammonites are abundant in thick shales which reach beach level at the point. Above the *Psiloceras*-bearing beds is a division composed mainly of shale and mudstone with a few limestone nodule horizons. These beds are referable to the *Alsatites liasicus* Zone and contain *Schlotheimia spp.* Many ammonites are visible in the foreshore reef exposures west of the promontory. Alternating thin limestones and shales which overlie the predominantly shaly *A. liasicus* Zone beds are referable to the *Schlotheimia angulata* Zone. Fossils are not common in these beds but *Schlotheimia sp.* occurs, as do specimens of the bivalve *Plagiostoma giganteum.* Palmer (1972) described the Lower Lias of the area.

Crossing the Doniford Stream brings the party to Doniford Bay where extensive offshore reefs in Lower Lias strata are developed.

Acknowledgments

The contribution of A. Whittaker is published by permission of the Director, Institute of Geological Sciences.

11
Aust Cliff

D. Hamilton

Objectives The purpose of the excursion is to examine the change from continental Keuper through Rhaetic to Liassic marine sediments, with their abundant fossils (including bone beds), sedimentary structures and minerals.

Special Features Low tides are essential for the excursion. Tides rise very rapidly along this section of coast (with a tidal range of about 12 m) and strong currents sweep past the bridge abutment. Visitors should begin the excursion on a falling tide. If caught on the east side of the bridge abutment when the tide rises, visitors should continue east along the coast to the end of the cliff where a path leads back to the motorway service area.

Maps

Ordnance Survey	One-Inch Sheet 155 (Bristol & Newport)
	1:50 000 Sheet 172 (Bristol & Bath)
	1:25 000 Sheets ST 58, ST 59
Geological Survey	One-Inch Bristol District Sheet

Main References Reynolds (1946); Whittard (1949).

Location Aust Cliff is situated at the eastern end of the Severn Road Bridge over the Severn Estuary (Fig. 1) at ST 566 898. Good general views of the cliff and rock platform may be obtained from the footpaths on the Severn Road Bridge, especially at low tide.

An access road (B 4461) leads to the south end of Aust Cliff from the A 403, which runs between the Severn Bridge junction on the M 4 (Junction 21) and Avonmouth. Go through the iron gates at the end of the access road (ST 564 889) and walk along the concrete causeway at the base of the cliff. By walking out on to the upper tidal flats of the estuary a more extensive view of the cliff is obtained.

Outline Geology This fine cliff exposure, first described by Buckland & Conybeare in 1824, shows:

1 The stratigraphic succession from Triassic Keuper to lowest Liassic strata in the cliff section.
2 Faulting in a broad anticline.
3 Folding at the base of the Keuper Marls.
4 Carboniferous Limestone and Dolomitic Conglomerate in the rock platform at the base of the cliff.

The Mesozoic succession exposed in the cliff (Fig. 2) is easily subdivided on the colour of the strata and on fossils. The lower part of the cliff is Triassic red Keuper Marl, followed by Tea Green Marls. Macrofossils are generally absent from these beds but occur abundantly in the overlying dark and lighter grey Rhaetic sediments. The prominent limestones and shales at the top of the cliff are Liassic in age.

The oldest rocks in the area are Carboniferous Limestones, which outcrop in the rock platform just upstream from the Severn Bridge. These formed part

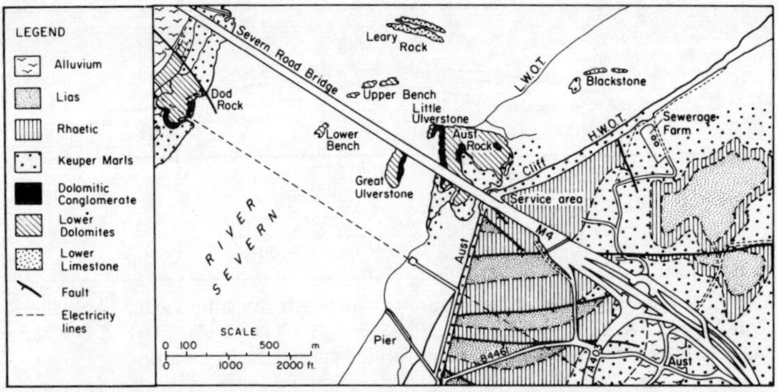

Fig. 1 Geological map of the Aust Cliff area (after Whittard 1949). Access to the cliff is from the south, along B 4461.

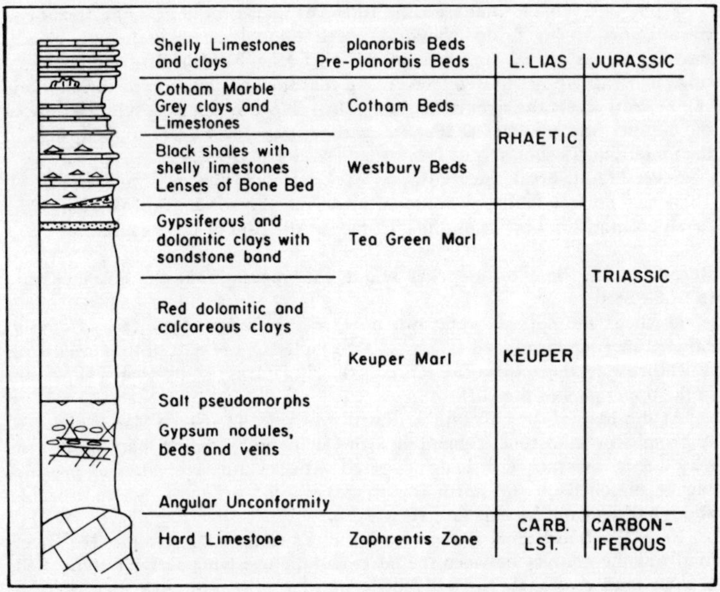

Fig. 2 The succession of Triassic to Jurassic rocks exposed in Aust Cliff.

of a Triassic landscape, in the valleys or wadis of which accumulated the Dolomitic Conglomerate. This is a limestone breccia, the matrix of which now consists largely of dolomite. Widespread deposition of sediments, in Upper Triassic Keuper times, buried the limestone topography. The Keuper Marls are red dolomitic and calcareous siltstones with gypsum and salt (halite) pseudomorphs, which were laid down in extensive playa lakes or on intertidal flats or enclosed bays. In the cliff face, note the irregular change in colour from the red Keuper sediments to the Tea Green Marl. The overlying grey Rhaetic sediments have distinctive hard limestone bands. The Lower Rhaetic Westbury Beds are dark grey in colour whilst the Upper Rhaetic Cotham Beds

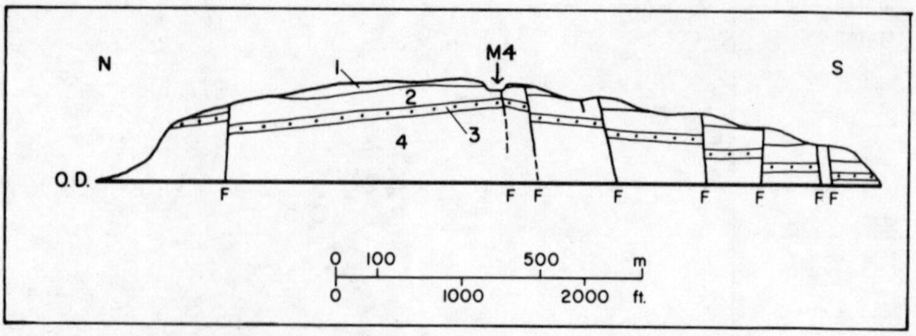

Fig. 3 The broad anticlinal structure of Aust Cliff (modified after Buckland & Conybeare 1824). 1 – Lias. 2 – Rhaetic. 3 – Tea Green Marl. 4 – Keuper Marl. M4 – Motorway cut on Severn Bridge approach. F – Fault.

are light grey. The topmost beds of the cliff are usually interbedded hard limestones and soft mudstones of Lower Liassic age. These are grey in colour when fresh but weather to light brown.

From this vantage point on the tidal salt marsh, note how the strata rise toward the Severn Bridge, beyond which they decrease in height, clearly showing the anticlinal structure of the cliff (Fig. 3). The SE–NW anticlinal ridge is breached by the River Severn, so that the continuation of the anticline can be seen across the river in Sedbury Cliff. The anticline is developed only in the Keuper and younger sediments as they drape over a flat-topped ridge in the underlying Carboniferous Limestone (Whittard 1949).

Several faults break the continuity of the strata in the anticline (see Fig. 1). These are normal faults, with downthrows to the south (downstream) side. Small promontories occur on the cliff face at the sites of these faults.

Itinerary Return to the base of the cliff about fifty paces from the entrance gate (ST 564 889).

Much of the cliff is overgrown now, so the sketch of the face as seen in earlier times is reproduced as Fig. 4. Two faults are present, both normal with downthrow to the south. The more northerly fault is plainly visible at present in the upper part of the cliff.

At the base of the cliff on the downthrow side, Tea Green Marl is exposed. A prominent sandstone, cemented with calcite and in which baryto-celestine may occur, together with beds peppered with small dark crystals of gypsum, can be matched on the north (upstream) side of the fault, where it occurs about half way up the cliff face (Fig. 4).

Note the transitional change from the red Keuper Marl to the Tea Green Marl but the contact between the latter and the overlying dark Westbury Beds is sharp and erosional. No Sully Beds are present here though they occur at this horizon on the western side of the Bristol Channel.

The lowest horizon of the dark Westbury Beds at this outcrop consists of a highly pyritized sandstone, only a few centimetres thick and often containing colourless tabular crystals and rosettes of selenite ($CaSO_4.2H_2O$). Elsewhere along the cliff, the Rhaetic Bone Bed occurs at the base of the Westbury Beds, where it forms lenses of grit and conglomerate, rich in vertebrate remains. Fallen blocks, from beyond the end of the concrete causeway, are the best source of specimens. (See p. 115 for a discussion of the origin of the Bone Bed.)

The basal sandstone passes up into isolated sandstone ripples in the overlying shale. Many of these sandstone ripples, which can be collected along the foot of the cliff, show abundant trace fossils indicative of a sub-littoral

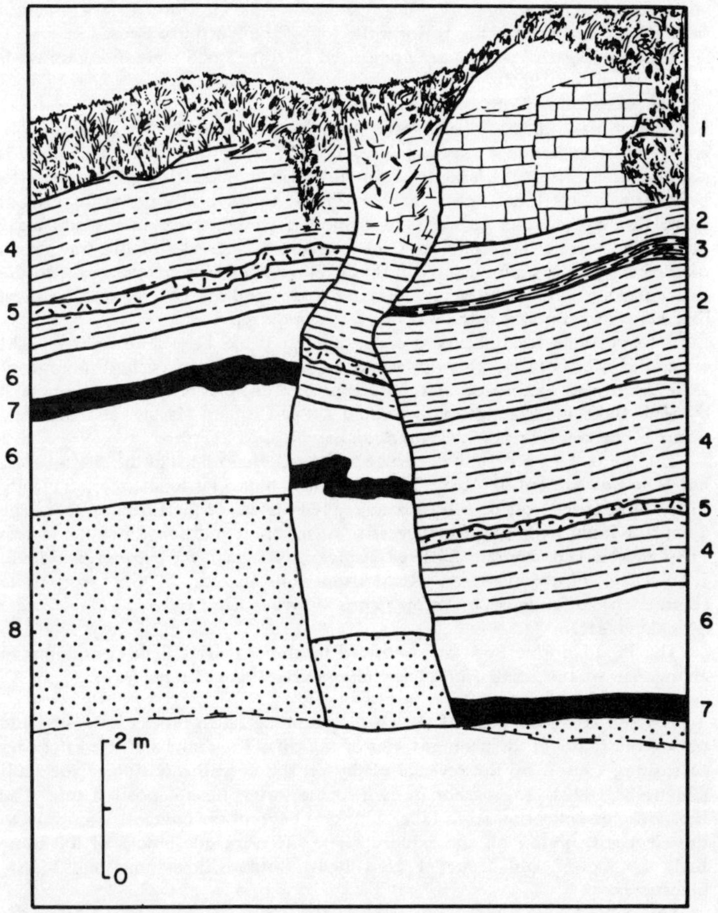

Fig. 4 Juxtaposed faults near the southern end of Aust Cliff. 1 – Lias. 2 & 3 – Cotham Beds. 4 & 5 – Westbury Beds. 6 & 7 – Tea Green Marl. 8 – Keuper Marl.

environment. These lowest Westbury Beds pass up into a fissile shale. The lower parts of these shales are fossiliferous, with *Rhaetavicula contorta* (Portlock), *Chlamys valoniensis* (Defrance), insect wings, beetle elytrae and small fish scales and teeth. Rosettes of selenite are common and the joint surfaces are frequently coloured reddish brown with jarosite.

The next prominent bed is the Lower Pecten Bed, a current concentrated sandy biosparite, with *Chlamys valoniensis* (Defrance), *Protocardia rhaetica* (Escher v.d. Linth) and other bivalves, together with fish scales and teeth. Generally, the shells have been dissolved away leaving only casts and moulds. Above the Pecten Bed, sand lenticles in the shale pass up into dark blocky shales. These have a sufficiently high hydro-carbon content to burn with a yellow smoky flame when heated fiercely.

The succession of a hard sandy shelly limestone, the Upper Pecten Bed with abundant *Pleurophorus elongatus* Moore, passing into a dark greenish shale is repeated once again, to form the top of the Westbury Beds.

Thus the Westbury Beds are comprised of three small scale fining upwards cycles (Hamilton 1962).

The succeeding Cotham Beds similarly have three argillaceous limestones which each pass up into clays. When fresh these Cotham Beds are blue grey in colour but weather to a very pale grey. Fossils are quite scarce in these beds and are of a different character from those in the underlying and overlying beds. The lowest limestone and shales appear to be locally unfossiliferous at Aust, though at other localities nearby derived Westbury fossils and algae occur. The second limestone contains sporadic occurrences of the tiny leaves of the liverwort *Naiadita lanceolata* Buckman, emend Harris, together with the ostracod *Euestheria minuta* var. *brodiana* Jones and insect remains. The third limestone also contains ostracods, e.g. *Darwinula* spp.

A fourth limestone occurs at some places but has been removed by slight erosion at other places. This uppermost limestone has impersistent mounds of algae, known as Landscape Marble (Hamilton 1961). It is often accompanied by mud flake breccia, which is termed Crazy Cotham Marble, and this may occur as channel fillings with *Pholidophorus higginsi* Egerton.

The White Lias generally occurs above the Cotham Beds in the Bristol area, but it is not present at Aust. Coarser grained Blue Lias limestones, typically shelly biosparites, rest directly on an eroded surface of the Cotham Beds. The abundant fossil assemblage is suggestive of marine conditions since the lowest Pre-planorbis beds contain *Liostrea hisingeri* (Nilsson), *Plagiostoma giganteum* J. Sowerby, *Pleuromya tatei* Richardson and Tutcher, *Atreta intusstriata* (Emmerich), *Pseudolimea hettangiensis* (Terquem), *Oxytoma longicostata* Strickland, etc.

The Pre-planorbis Beds are followed by the planorbis Beds containing an abundance of the small *Hettangian ammonite, Psiloceras planorbis* (J. de C. Sowerby).

Proceed along the base of the cliff, examining fallen blocks for fossils and noting the faults at the promontories of the cliff. The faults are often the sites of springs, caused by the reversal of dip on the downthrow side of the fault (Reynolds 1946). The carbonate rich spring waters have deposited tufa. The best area for collecting from fallen blocks is beyond the concrete causeway to the electricity pylon on the estuary; these often include blocks of the Bone Bed, the Lower and Upper Pecten Beds, Cotham limestones and Liassic limestones.

On approaching the Severn Bridge, the lower beds of the Keuper are exposed, where gypsum nodules and alabaster veins occur at the base of the cliff. The gypsum is probably derived from the alteration of original anhydrite and has later migrated into joints. Whittard (1949) suggested the polygonal pattern may be due to desiccation cracks but the joints are more probably related to the slight folding. At the foot of the cliff beyond the bridge, cuboidal salt pseudomorphs occur in greyish siltstones. These have formed on the underside of beds by the infilling of cavities formed by the solution of halite crystals. A small anticline is visible in the lower part of the cliff but this is quite local. If tidal conditions are favourable, an anticlinal structure can be traced in Keuper sediments on the rock platform. As these structures do not occur in the underlying Carboniferous Limestone, Whittard (1949) suggests that the structures in the cliff are due to compaction over a ridge of Carboniferous Limestone, which is of Z zone or Black Rock Limestone age.

Discussion

Having examined the whole section some further points concerning mainly the environments of deposition will be discussed.

1 **Keuper Sediments** The red and green Triassic sediments are fine-grained,

of predominantly silt grade and have been laid down in water. The siltstones are moderately hard when dry. When placed in water some of these break down very rapidly, due to the presence of expanding chlorite clay minerals. Other specimens may not break down on immersion due to the presence of calcite, dolomite or even gypsum cements. The presence of gypsum, pseudomorphs of halite and the widespread occurrence of dolomite and calcite suggests deposition in hypersaline lake conditions. When sandgrains of medium size occur in these sediments, the grains are very well rounded, suggesting aeolian abrasion under desert conditions.

The boundary between the red Keuper siltstones and the Tea Green Marls is based on the colour difference. In detail this boundary is most irregular for green patches commonly occur within the red, and less commonly red patches occur within the green country rock. The origin of these colour differences are still a matter of controversy. The red colour is due to finely divided ferric oxides (hematite and goethite) absorbed on to the fine sediment particles, whilst the green is due to the absorption of ferrous oxides. Typical analyses yield between 3–4% of Fe_2O_3 and less than 1% of FeO in the red siltstones, whilst in the green siltstones, less than 1% Fe_2O_3 and more than 1% FeO are present. Thus the change from red to green may be accompanied by a reduction in the total iron present, suggesting that solution and removal of some iron has occurred with the change in the state of oxidation. The ferrous state appears remarkably persistent in atmospheric conditions so the iron oxides are probably strongly adsorbed on the sedimentary particles.

Though the red colour appears to give rise to the green, it remains uncertain as to the time when these sediments first became red in colour. They could have been red when deposited, suggesting an origin from a lateritic source, or oxidised during or shortly after deposition, or oxidised long after the deposition of the sediments. The change from red to green often occurs along open joints but is most pronounced adjacent to the highly organic and pyritiferous Westbury Beds. It seems probable that this extensive change in colour at the top of the Keuper Marls is not related to the original depositional conditions of the Tea Green Marl but to reduction of the oxides in the low pH and Eh conditions induced following burial by the overlying Westbury Black shales.

2 **Rhaetic Sediments** *Rhaetic Bone Bed* The well known bone bed consists of a conglomerate or breccia of sedimentary rocks in a calcite cemented sandy matrix, together with an abundance of bones, teeth and scales. Near Aust this bed occurs in patches at the base of the Westbury Beds. Abundant vertebrate remains are common in the succeeding shelly limestones of the Westbury Beds and scattered vertebrate remains occur also in the shales. The term Rhaetic Bone Bed normally applies to the conglomeratic phase.

This occurs directly on top of the Tea Green Marl, the surface of which may be rippled. The conglomerate at Aust is made up predominantly of lumps of Tea Green Marl together with quartz pebbles. The Tea Green Marl forms rounded, apparently rolled masses, but the presence of squeezed and plastically deformed masses, shows that the Tea Green Marl was soft and of very local derivation. A possible source could be from low cliffs (only a metre or so high) cut by wave action as the Rhaetic transgression progressed across the landscape of low relief. In this way, locally derived sediments would be the major contributors to the conglomeratic beds. The quartz pebbles are mainly of vein quartz and are mostly well rounded. They could be derived from older beds (polycyclic) or they could be the stomach stones (gastroliths) swallowed probably by *Pleisiosaurs* to aid the digestion of their food (Wickes 1908).

The major interest in the bone bed is, of course, the abundance and variety of vertebrate skeletal remains. In addition, coprolites (faecal droppings), especially of aquatic reptiles, are very common; these contain crustacean fragments and abundant fish scales. Though jaw-bones with teeth have been found, the remains are usually disarticulated and frequently show signs of abrasion. The vertebrate remains comprise fish and reptiles (see Reynolds 1946). Three groups of fishes are present in the fossil remains, elasmobranchs, ganoids and dipnoans.

The teeth of the primitive elasmorbranch sharks *Acrodus* and *Hybodus* are very common. The former consists of a small low dome whilst the latter has a principal cone with smaller lateral cones. Both these sharks had powerful fin spines, which may reach up to 30 cm in length. A smaller fin spine, with only a few large denticles, is referred to *Nemacanthus monilifer* Agassiz. The chondrostean ganoids are primitive bony fish. Sharply conical teeth, having an enamel cap and a prominent collar, assigned to *Bigeria acuminata* (Agassiz), are common in the Bone Bed. Teeth with chisel or knob-like crowns and long almost cylindrical roots are termed *Sargodon:* the numerous small rhomboidal scales termed *Gyrolepis* probably belong to the same fishes. The most common remains of the dipnoan lung-fish are the stout multi-cuspate dental plates of *Ceratodus latissimus* Agassiz. Present-day *Neoceratodus* lives in rivers and swamps of Australia.

The reptiles are mostly remains of ichthyosaurs and plesiosaurs, with in addition probably some dinosaurs. Remains have often been identified to specific level but the quality of the material does not justify such precision. The most frequently preserved bones of the fish-shaped ichthyosaurs are discoid vertebral centra, which have two concave surfaces. Both the teeth and vertebrae of *pleisiosaurs* are very common, the vertebrae being distinguished from those of *ichthyosaurs* by, among other features, being thicker and having two planar surfaces on the centra. Some very large bones have been referred to the terrestrial carnivorous dinosaur *Avalonia.*

Thus the vertebrates represented in the Rhaetic Bone Bed are predominantly aquatic fish and reptiles with some terrestrial (?littoral) dinosaurs, at least some of which could have been fresh or brackish water species. No amphibian labyrinthodonts are known (Savage & Large 1966). A similar fish fauna, though less abundant than in the conglomeratic phase and without the reptiles, occurs in the basal sands of the Westbury Beds, and also in the base of the limestone bands and in the shales. Thus the Bone Bed is a concentration of the Westbury Bed vertebrate fauna.

Many hypotheses have been advanced to account for the abundance of vertebrate remains in the main bone bed. These include mass mortality due to the effect of high salinity as marine animals entered the hypersaline Keuper lakes (Jukes-Browne 1892); storm deposits on coastal flats (Short 1904); concentration of vertebrate remains in aquatic feeding grounds (Wickes 1904); concentration due to a reduced rate of deposition of sediment (Richardson 1901); wave action (Donovan 1955).

The basal conglomeratic bone bed is suggestive of a strand line deposit with wave action and shore line currents winnowing and abrading the material. Some of these vertebrate remains could have been derived through reworking of Keuper sediments.

Cycles of Sedimentation Both the Westbury Beds and the Cotham Beds are characterized by the sequence of a sandy or shelly limestone resting on an erosion surface; the rapid passage upwards into shales with ripple lenticles of sand; a decrease upwards in the amount of sand, as the sand lenticles give way to sand laminae a few grains thick; and an uppermost portion of laminated or blocky mudstones. There is an obvious decrease from the bottom to top in the energy level of the depositional environment of these small scale fining upwards cycles (Hamilton 1962).

The basal beds are current concentrated sands or more generally shell beds, which in the Westbury Beds are termed the Lower and Upper Pecten Beds.

Current activity is shown by the orientation of the shells, which are almost all convex upwards and alignment of elongated shells such as *Pleurophorus elongatus* Moore. Though these shelly limestones occur over widespread areas, they wedge laterally, suggesting the possibility of their being sub-aqueous migrating channel deposits or sub-littoral wave concentrated deposits. The shells comprising these beds, such as *Rhaetavicula contorta* (Portlock), *Chlamys valoniensis* (Defrance), have been transported and are not shell banks.

The sand ripple lenticles retain a profusion of trace fossils on their upper and lower sides and these are best collected along the front of the cliff. The ripple lenticles formed under conditions of decreasing currents and availability of sand, such as would occur with an increase in depth. The upper mudstones could represent the deeper water phase of deposition, for the fauna is thin shelled and fragile.

Faunas of the Westbury Beds and Cotham Beds The fauna of the Cotham Beds is quite distinctive from that of the Westbury Beds. The fauna of the Westbury Beds has a relatively small number of molluscan species but these occur in quite large numbers, as is typical of brackish water assemblages. As well as the aquatic fish and reptiles already noted, ophiuroids (brittle-stars) and the brachiopod *Orbiculoidia townshendi* (Davidson) are recorded. The presence of these latter two suggests that there was an opening to marine conditions.

In contrast the main constituents of the Cotham Bed faunas are: the liverwort *Naidaita lanceolata* Buckman, with an associated microflora, which was interpreted as a lake environment by Harris (1938); an abundance at some horizons of the ostracod *Darwinula* spp.; the occurrence of the phyllopod (arthropoda) *Euestheria minuta* var. *brodiana* Jones; algal mounds, giving rise to the growth forms of Landscape Marble. These are consistent with fluctuating lacustrine conditions.

3 **Liassic Sediments** The shallow water shelly limestone resting on the eroded top of the Cotham Beds heralds the marine incursion of the Lias. There is a diverse molluscan fauna of marine aspect, which includes echinoids and corals. Truly marine conditions are established by the time the ammonite *Psiloceras planorbis* is present.

Thus the changes from Triassic terrestrial deposition, through the fluctuating Rhaetic transgression to the Liassic marine transgression are recorded in the succession from the base to the top of Aust Cliff.

Minerals

Though not a good collecting site for mineral specimens, the following may sometimes be found in fallen blocks along the cliff.

Baryto-celestine/$(BaSr)SO_4$. Varying amounts of barium and strontium may be present in this sulphate. Occurs very rarely in the Tea Green Marl, usually as very pale blue elongate crystals radiating fan-wise on joint planes.

Calcite/$CaCO_3$. Coarse, creamy to white crystals sometimes occur in veins, especially in the Lower Liassic limestones. Dark prismatic calcite often forms at the bottom or top of the Westbury limestones. The elongated crystals are arranged in horizontal bands at right angles to the bedding and are sometimes called 'beef'.

Galena/PbS. Small silvery cubes of this lead sulphide have been noted extremely rarely in the Bone Bed (Hamilton 1966).

Gypsum/$CaSO_4.2H_2O$. This may occur in several forms. Dull, earthy rock gypsum occurs as nodules in the Keuper Marl just downstream from the abutment of the Severn Road Bridge. Here also, pearly, fibrous satin spar, and massive white alabaster can be found. Near the base of the Westbury Beds, small colourless, blade shaped crystals occur in joints. These are the variety selenite and they generally form small rosettes.

Jarosite/$K_2Fe_6(OH)_{12}(SO_4)_4$. Occurs as a brownish surface incrustation in the Bone Bed and the lower Westbury shales.

Pyrite/FeS_2. Occurs most frequently on the upper and lower surfaces of the Westbury limestones and sand lenticles.

Salt pseudomorphs. These are not truly minerals but are formed by the replacement of salt (halite, NaCl) cubes by sediment as the salt dissolved away. The cuboidal shape of the salt crystal is preserved in the sediment, usually on the underneath side. Salt pseudomorphs may be found occasionally in green bands in the Keuper Marl, just upstream of the abutment of the Severn Road Bridge and further east as the end of the cliff.

12
Jurassic rocks of the Bath area

A. B. Hawkins

Objectives

Although William Smith described Rugborne Farm, High Littleton as the birthplace of geology, he suggested Bath could be considered the cradle of geology. This excursion will offer the visitor the opportunity to examine the various lithologies in the Lower and Middle Jurassic, concentrating mainly on the latter. Where appropriate brief reference will be made to the superficial structures, which affect the outcrop thickness, the near surface dips and the vertical position of the surface strata relative to the height of the *in situ* horizon.

Special Features

Some of the roads to localities are unsuitable for coaches (see Fig. 1). Permission is required to visit the following localities:

Fuller's Earth Works: The Manager, Midford Works, South Stoke, Bath. Combe Down 833343.

University of Bath: Gardens Superintendent, The University, Bath. Bath 6941.

Beacon Hill: Messrs M. P. Kent Ltd., The Towers, Beacon Hill, Bath. Bath 64184.

Maps

Ordnance Survey	One-Inch Sheets 156 (Bristol & Stroud), 166 (Frome)
	1:50 000 Sheet 172 (Bristol & Bath)
	1:25 000 Sheets ST 75, ST 76
Geological Survey	One-Inch Sheets 265 (Bath), 281 (Frome)

Main References

Cox *et al.* (1941); Green & Donovan (1969).

Outline Geology

The oldest formation that outcrops in the Bath area is in the Pennant Series of the Upper Coal Measures, which forms an inlier in the Newton St. Loe area (ST 700 650) Table 1. When observed in temporary exposures this can be seen to be unconformably overlain by the Keuper Marl. Frequently the crop of the Keuper Marl is veneered by several metres of foundered Liassic debris. The overlying Rhaetic beds are only seen in temporary exposures. The limestone-rich part of the Blue Lias frequently forms rock benches, yet at present the best exposure is in a valley at Pennyquick Bridge (loc. **1**). Unfortunately the overlying Lower Lias Clay, Dyrham Silts (where present) and the Junction Bed are rarely exposed and nowhere permanently. The Midford Sand of the Upper Lias can be seen in partially overgrown outcrops at Midford, Beechen Cliff and Beacon Hill (loc. **8**, **10**, **11**).

Although on faunal evidence the Lower and Middle Inferior Oolite are missing, there is no clear angular unconformity between the Midford Sands and the Upper Inferior Oolite (loc. **7**, **11**). The basal bed of the Upper Trigonia Grit is conglomeratic (Richardson 1907, 1909) containing gravel-sized clasts of quartz and very fine red sandstone as well as rounded cobbles of calcareous sandy siltstone, the latter probably representing reworked cemented Midford Sands. The Upper Trigonia Grit is frequently very fossiliferous with bands of

JURASSIC	Forest Marble	Forest Marble Clays and Limestones
		Hinton Sands
		Forest Marble Limestone & Clays
	Bradford Clay	
	Great Oolite	Upper Rags
		Bath Oolite
		Twinhoe Beds
		Combe Down Oolite
	Fuller's Earth	Upper Fuller's Earth Clay
		Fuller's Earth Rock
		Lower Fuller's Earth Clay
	Inferior Oolite	Upper Inferior Oolite
		Rubbly Bed
		Anabacia Limestone
		Doulting Stone
		Upper Coral Bed
		Upper Trigonia Grit
		(Middle & Lower Inferior Oolite missing)
	Upper Lias	Midford Sands
		Junction Bed
	Middle Lias	Dyrham Silts (where present)
	Lower Lias	Lower Lias Clays
		Blue Lias
TRIASSIC	Rhaetic	White Lias
		Cotham Beds
		Westbury Beds
	Keuper Marl	
CARBONIFEROUS	Coal Measures	Pennant Series

Table 1 The succession in the Bath area

silty and sandy biomicrudite interbedded with pelmicrite. According to Richardson (1907) the Dundry Freestone occurs at Englishcombe but does not extend to the Midford area. The Upper Coral Bed is exposed throughout the area, being thickest (4.3 m) between Midford and Combe Hay. This bed generally consists of *Isastrea* colonies and nodules of micrite and biomicrite set in a marl matrix. However, in the Combe Hay to Midford area the top 0.5 m is generally a biosparite, occasionally oosparite, with a waterworn oyster-covered upper surface. The Doulting Stone is dominantly a thickly bedded oosparite although bands of biosparite and pelsparite are present. Outer moulds of terebratulids and *Trigonia* sp. occur, especially towards the top. As the basal bed of the Anabacia Limestone is also an oosparite the junction is not always clear. However, much of the Anabacia Limestone is a harder creamy pelmicrite often with closely spaced vertical discontinuities which clearly differentiates the two beds when weathered. The small button-shaped coral *Anabacia complanata* (Defrance) occurs both in the Doulting Stone and the Anabacia Limestone; hence it is not possible to use the presence of this fossil to indicate the Anabacia Limestone. The thin Rubbly Beds are not always seen. Generally they are a biomicrite containing abundant terebratulids and bivalves.

The Fuller's Earth Clays in the area were described by Arkell & Donovan (1952), Torrens (1968), and mentioned by Green & Donovan (1969) (loc. **3**, **13**). The Lower Fuller's Earth Clay is usually a calcareous fissured mudstone weathering to clay near the surface. Near the top of the bed is the very fossiliferous band with *Liostrea acuminata* (J. Sowerby). The simple tripartite lithological subdivision of the Bath area is complicated to the north where the Fuller's Earth Rock splits into three limestone bands (Sheet 265, Torrens 1968). In the south the Upper Fuller's Earth Clay is thicker and contains the commercial Fuller's Earth Seam approximately 4–8 m below the base of the Great Oolite in the Combe Hay area. If, as suggested by Hallam & Sellwood (1968), the seam has a volcanic origin then it may be regarded as a time constant horizon. Unfortunately, however, it does not exist to the north of Bath so it is of very restricted use for correlation.

Green & Donovan (1969) have recently published a revised classification of the Great Oolite of the Bath area (loc. **2**, **4**, **6**, **9**, **12**). This begins with the Combe Down Oolite, generally a current-bedded oosparite and biosparrudite. The top of the Combe Down Oolite is frequently a planed, bored and often oyster-covered surface. The overlying, frequently ferruginous Twinhoe Beds have been sub-divided by Green & Donovan (1969) into three facies. Above is the celebrated Bath Stone, dominantly oosparite but sometimes containing oomicrites and biosparrudites. The succeeding Upper Rags are a variable sequence dominantly of biosparrudites and oosparites with considerable lateral variation including the Corsham and Bradford Coral Beds in the respective areas.

The Forest Marble is generally a silty mudstone but arenaceous (loc. **5**) and carbonate facies (loc. **12**) are locally present. In the latter case it is often difficult on lithological grounds to separate the oosparites and biosparrudites from those of the Upper Rags of the Great Oolite.

The gentle easterly dip of the Jurassic rock in the Cotswolds is modified by the broad upfold in the Kelston area and by the many superficial movements. The upfold is responsible for raising the level of the Blue Lias limestones to approximately 70 m in the Kelston area whereas in Bath itself this bed is below the level of the alluvium (20 m). Because of the interbedded limestones/mudstones of the Jurassic rocks of the area, the relief amplitude of over 200 m and the periglacial condition during the Quaternary, the surface geology of this area has suffered extensively from superficial movements. The dominantly sub-parallel outcrops on the old series geological maps (1873) have been replaced on the new Bath Sheet (265) by 45 km^2 of foundered and 20 km^2 of landslip strata. Unfortunately by implication this suggests the remaining strata are not disturbed, something frequently not true. The most important camber slopes are in the Inferior Oolite. Many of the 'draps' in the Englishcombe area (loc. **2**) are indicated as camber slopes on the One-Inch geological maps but similar 'draps' near Tog Hill and east of Cold Ashton are not marked. Only in temporary exposures is it possible to see the typical superficial features in the Inferior Oolite. However, although cambering is not so well developed in the Great Oolite, exposures are more frequent and hence gulls, dip and fault structure, etc. can be more easily seen (loc. **2**, **6**, **9**). The effects of big landslides, especially those of Beacon Hill and Beechen Cliff, can be clearly seen because of the slopes produced by their backwalls (loc. **10**, **11**). Smaller slab slides still occur, especially on the Fuller's Earth mudstones (loc. **4**, **18**, **13**).

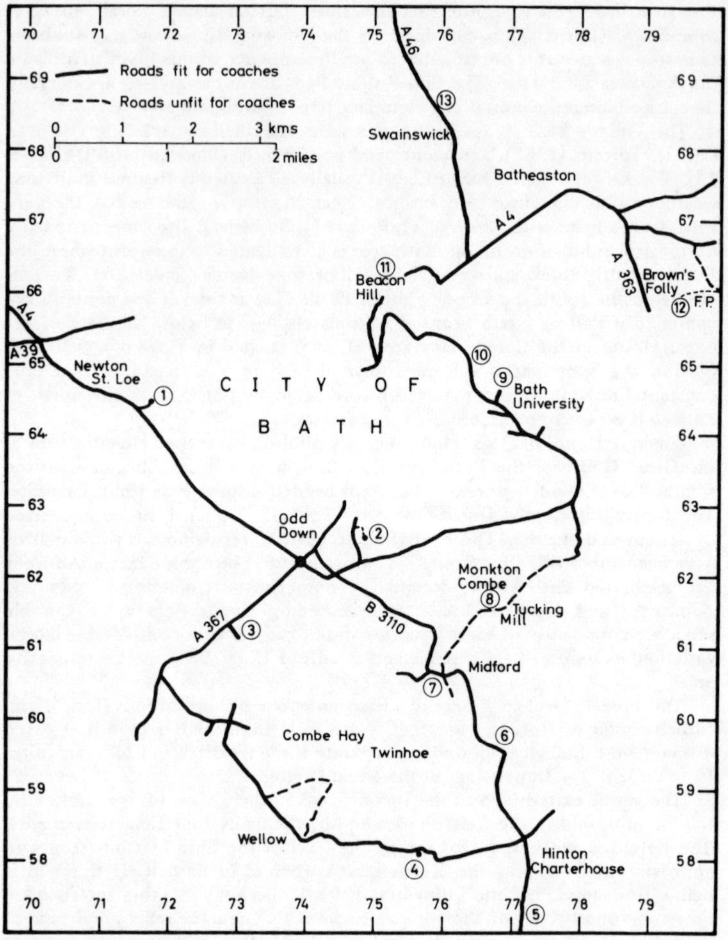

Fig. 1 Locality map

Itinerary

1 Pennyquick Bridge (Blue Lias) (ST 716 645) From Bristol proceed along the A 4 towards Bath. At the traffic lights near the Globe Inn at Newton St. Loe turn right, then immediately left, taking the road variously signposted; Odd Down being the most prominent of the places mentioned. In the spring or autumn when the fields to the north and east of the inn have been freshly ploughed, black spoil heaps on the red-brown Keuper Marl indicate that the Pennant Series in this area was once mined.

After crossing the Newton Brook about 1.5 km along this road, turn left and proceed to the exposure about 150 m up the road. This previously undescribed section of the Blue Lias is now the best permanent exposure of this formation in the Bath Area, exposing just over 3 m of interbedded limestone and thinly laminated shelly mudstones. The limestones make up about 70% of the deposit, the thickest band being 0.35 m. Waterworn *Arnioceras* sp. and also *Coroniceras* (*Primarietites*) *revnesi* (Spath) have been obtained. It is probable, therefore, that the beds exposed represent the *conybeari* – *revnesi* subzones seen at Saltford (Donovan 1955) and Keynsham (Hawkins 1966). The exposed upper surface of the fourth limestone band has many outer moulds of ammonites and bivalves including *Liostrea hisingeri* (Nilsson) and *Gryphaea* sp. *Calcirhynchia calcaria* (Buckman) can be found in the upper limestone bands, while the large bivalve *Plagiostoma giganteum* (J. Sowerby) is present, mainly in the lower limestone bands. The bases of many of the limestone bands are very irregular, the protuberances being up to 0.15 m deep. These features are probably early post-depositional structures similar to those described by Webber (1967) in South Wales.

Calcite crystals are seen along master joints in the upper limestone bands. The dip of the beds, up to 10°, is probably associated with superficial structures or possibly drag due to the effect of the Pennyquick Fault which truncates the Blue Lias on the south side. Unfortunately the downfaulted Lower Lias Clay is now obscured by vegetation.

2 **Springfield Quarry** (Great Oolite) (ST 748 625) Rejoining the Odd Down road, proceed for 1.5 km, stopping just past Twerton Hill in the vicinity of the Jubilee Inn where, to the right, a good view can be obtained of the slopes in the Newton Brook area. To the south of the brook the long 3–4° slopes are developed on the Inferior Oolite. In the Nailwell area the standard 12–13 m thickness can be measured, yet in the slopes at Inglesbatch the strata are mapped by the Institute of Geological Sciences over a height exceeding 60 m. To the north of the brook, in the Wilmington area, the camber 'drap' exceeds 75 m. Proceed for another 2 km to the roundabout at the junction with the A 367 and pass straight across, along Frome Road for a further 0.5 km to the junction with Midford Road. Continue straight across into Bradford Road. Take the second left hand turning into Entry Hill and, after 0.5 km, turn right into Entry Hill Park. A further right turn will bring you to the quarry entrance. The quarry is now a public park administered by the Parks Department of Bath Corporation. Hammers must not be used and visitors should not climb the quarry faces.

The strata exposed here are the Combe Down Oolite overlain by the Twinhoe Beds. The Combe Down Oolite, up to 2.25 m thick, consists of thinly to medium bedded oosparites with occasional bands containing sufficient comminuted shells to be described as biosparrudites. The large scale foresets dip at 10–30° in a southerly direction. The top of the Combe Down Oolite is a planed, bored surface which in parts is oyster-covered.

The Twinhoe Beds begin with an impersistent 0.1 m marl bed which is overlain by up to 4 m of variable deposits. The lowest 1 m is generally a biosparrudite with abundant unorientated, comminuted shells of rhynchonellids, terebratulids and oysters. Sometimes solution has removed the whole shells giving a cavernous appearance to the yellowish-cream rock. The whole of the layer is current-bedded but in places, especially in the lower half, the cross-bedding is fine and sometimes herringbone. Towards the top of the layer the shells appear less abundant and the deposit becomes an oosparite with bioclasts and pisoliths and with thin bands of oomicrite. In the southern end of the quarry this layer is seen to be overlain by another impersistent 0.1 m marl band. Above this is a 3 m thick yellowish-orange, slightly ferruginous, oomicrite with some bioclastic material and pisoliths. The top of the bed is planed and bored. It is overlain by irregularly bedded, light cream pisolitic biomicrudite.

This quarry shows some gulls, but the ones visible now are not so impressive as those covered over during the building of the estate. The concrete block walls cover the entrance to old adits into the Combe Down Oolite.

3 **Fuller's Earth Works** (Fuller's Earth Bed) (ST 729 612) From Springfield Quarry it is necessary to return up Entry Hill and along Bradford

and Frome Roads as far as the roundabout with the A 367. Taking the left-hand road towards Radstock proceed for 1.5 km before turning left off the straight main road into the Fuller's Earth Works. Here the mudstones of the Fuller's Earth seam may be examined as they arrive from the underground workings. In hand specimens the moist montmorillonite-rich mudstone of the seam has a slight greenish tinge, but otherwise is little different from the normal unweathered grey mudstones of the Fuller's Earth Clay.

Approximately 620 tonnes of Fuller's Earth Bed are mined each week. The seam is 18–25 m below the surface (Highley 1972). NW of the road the seam is about 2 m thick but east of the works it reaches 3 m. In general it is the lower 1 m which has the high montmorillonite (c.80%) and low calcite content (12–18%) – the 'Good fuller's earth' of Lonsdale (1835). The bed has a moisture content of 27% when mined and is dried to about 6% for the foundry and pharmaceutical industries. The upper part of the seam has a higher (20–45%) calcite content and consequently a lower montmorillonite percentage (c.50–75%) – the 'Bad fuller's earth' of Lonsdale (1835). This upper part is dried to about 3% moisture and sold as a carrier for fertilizers and as a binder in cattle food (Highley 1972).

The Fuller's Earth bed is mined by the pillar and stall method. When an area has been worked, the pulling of some of the supports causes roof collapse. The resulting subsidence at the surface, frequently within three months of working, can be seen in the fields to the west of the works. The E–W fault shown on Sheet 281 has about a 10 m downthrow to the NW. Although this obviously hinders mining development, it has no surface expression.

4 Hinton Hill (Great Oolite-Twinhoe Ironshot) (ST 757 582) Leaving the Fuller's Earth Works, continue SW along the A 367 for a further 1 km. Turn left opposite the Crossways House, along the clearly signposted Combe Hay Road. Unfortunately, during 1974 the good exposures in the Inferior Oolite along the old railway line between Combe Hay and Midford were infilled or spoilt by tipping and are now not worth visiting. Therefore, instead of going as far as Combe Hay, take the second right turning to Wellow. After crossing the Cam Brook and reaching the top of the ridge, those with cars may take the left hand turning towards Twinhoe, to the type locality of the Twinhoe Ironshot facies (ST 740 592). Here infilling and overgrowth make it very difficult to examine the section, although it is frequently quoted in the literature (Cox *et al.* 1941; Donovan 1964; Green & Donovan 1969). (See Excursion 13, loc. 6). Those following this route should take the next right turn down into Wellow, and there the left hand road to Hinton Charterhouse.

In winter it may be possible to see the Twinhoe Ironshot facies at Hinton Hill (ST 757 582). This now degraded exposure in the roadside woods, shows up to 2 m of the bed. Green & Donovan (1969) published photomicrographs of the typical Ironshot beds and also gave a long list of fossils from them. The rock is a biomicrite with ferruginised pisoliths, which, under the microscope are seen to be rounded intraclasts of oomicrite and biomicrite with a layered, probably algal coating. At this locality it is worth looking at the hummocky topography of the field to the north. Cracks are sometimes visible in the road indicating the active land-slipping which is taking place at the top of the Fuller's Earth sequence.

5 Hinton Sand Pit (Hinton Sands) (ST 772 573) Continue towards Hinton Charterhouse before turning right and proceed 1 km along the B 3110 towards Norton St. Philip. The sand pit is seen immediately to the left of the road, as it rounds a slight bend.

This disused sand pit is the type locality of the Hinton Sands, a localised facies in the Forest Marble. Here about 8 m of buff, slightly silty, fine grained sands occur, but only the lowest 3 m are now visible. Within the sands are large spheroidal sandrock bodies, the so-called doggers, which are up to 1.2 m maximum thickness and up to 3.5 m long. These irregular shaped calcareous concretions contain layers and lenses of unindurated sands. On the west face of the biggest dogger, signs of thin bedding can be seen in places.

Beneath the big dogger the loose sands contain layers of clay lenses generally 2–5 mm thick and up to 2.5 cm long. These occur dominantly in a band about 0.4 m below the dogger. Below this the colour banding suggests the possible presence of bedding, or past groundwater levels. Cox *et al.* (1941) record fragments of lignite, annelid tracks and a scanty bivalve fauna from the layer below the large dogger. They also point out that at the top of the quarry loose slabs of calcareous sandstone showed *Ostrea* sp. on the upper surface.

6 **Midford Hill** (Great Oolite) (ST 768 596) Leaving the sand pit, return towards Bath along the B 3110 for 2.5 km. It is advisable to park on the left hand side of the road just before the sharp bend. The disused and now very overgrown quarry is in the wooded area to the north of the road just round the bend and 20 m west of the lane to a field. It shows Upper Rags overlying Bath Oolite and was briefly described by Green & Donovan (1969).

As the quarry is normally entered at the top, the beds will be described in descending order. For about 1 m below the topsoil are thinly to very thinly bedded oosparites with occasional biosparrudites. For 0.95 m below the rock is similar but the relative organic content increases to produce 3–12 cm thick horizons dominantly of biosparrudite with intraclasts. In hand specimens some of the biosparrudites can be seen to be rich in polyzoans, while occasional pisoliths also occur. These rocks are cross-bedded, with foreset dips up to 23° southwards. To the west, in the second part of the quarry, the next bed (0.2 to 0.4 m thick) is easy to distinguish by its herring-bone cross-bedding. It is a biosparrudite containing pisoliths and polyzoans, frequently with a bored upper surface. At the base is a very thin (10 mm) band of oolite marl, below which is a 0.25 m bed of biosparrudite with polyzoans. Next comes the 2.45 m thick bed of oosparite and biosparite with some pisoliths and polyzoans. Below the thick bed at the next quarry level is a harder band of biopelsparite, 0.26 m thick. Towards the base of this band the matrix becomes micritic, pisoliths occur and in places there is a transition with the 0.44 m band of pisolitic marl with clay lenses beneath. The surface of the bed beneath the marl is frequently bored and in places cut by channels up to 13 cm deep, both being infilled with marl similar to the bed above. The basal bed, exposing 2.8 m of Bath Stone, is a massive oosparite and biosparrudite with pisoliths throughout.

The western end of this quarry shows the early effect of cambering, resulting in gulling of the Great Oolite. This is more apparent in the massive beds where the pre-Quaternary joints have been opened by tension, while in the thinly bedded upper horizon small-scale bed movement has resulted in the gulls not reaching the surface. On the surface of some rock, calcareous dripstone shows sub-horizontal calcite ridges. These probably accumulated before the rock was cambered and represent deposits formed at previous varying groundwater levels.

7 **Midford** (Inferior Oolite) (ST 761 607) Proceed northwards down Midford Hill to Midford. Here there are several small exposures which together show the succession through the Inferior Oolite. This locality was described by Richardson (1907, 1909) and remains the best recorded section of Inferior Oolite in the Bath area. Unfortunately, his 1907 description preceded the excavation of the Camerton to Limpley Stoke railway line (1907–10) hence it does not record the detail that was available only a few years later. The present author has noted many exposures along this old railway but these have now been generally infilled and hence are not recorded in this itinerary.

The detailed section given by Richardson (1907) and shown in Plate IV (1909), is 150 m along the Wellow Road south of the village (ST 761 604).

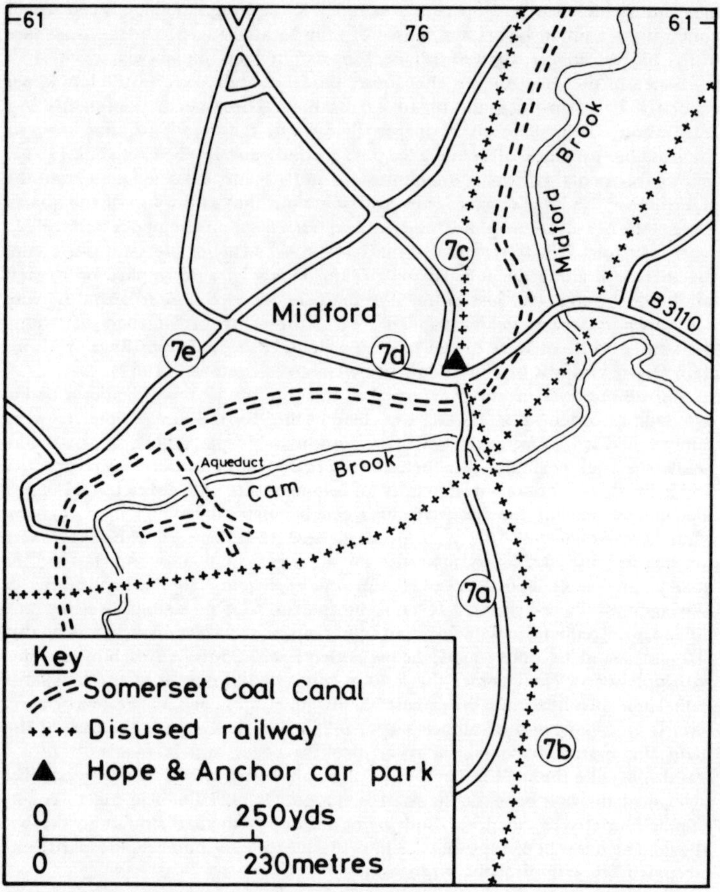

Fig. 2 Location of sites around Midford (loc. 7)

After only 20 m however, it is worth noticing that the gardens alongside the iron bridge occupy the position of the old Somerset Coal Canal, (see Excursion 13, loc. 8). The Oolite exposure (Fig. 2, 7a) is now much degraded and obscured by vegetation but it is possible to work out the bed description given by Richardson (1907). The Upper Trigonia Grit, poorly exposed by the pathway into a house (High Pines) some 120 m from the road junction, consists of slightly silty biomicrudite interbedded with a slightly ferruginous pelmicrite. The base is not seen. Richardson (1907, 1909) records a large number of fossils obtained when the face was fresh. The Upper Coral Bed, indicated as 2.45 m thick by Richardson (1907), is very poorly exposed and although lumps of *Isastrea* can still be obtained, it is better to collect from other nearby exposures.

The Doulting Stone exposed can be compared with that described by Richardson. Much of the 1.1 m lower band is a pelsparite with occasional intraclasts. Serpulid, *Trigonia* sp. and terebratulid moulds can be seen in the face. According to Richardson, the upper surface is waterworn and bored by lithophagi. The middle bed consists dominantly of an oobiosparite, again with a bored upper surface. The massive upper bed (1.35 m) is easily distinguished, and consists of an oosparite with some bioclasts, intraclasts and pellets. Locally moulds of terebratulids give the rock a cavernous appearance.

At present it is not possible to see the Anabacia Limestone in the road cutting but entering the old railway cutting across the road and proceeding southwards (7b), the Limestone is seen to be over 3.5 m thick, although the base is not visible. The lower part of the Anabacia Limestone is an oosparite with pisoliths but the dominant lithology is a pelmicrite. *Anabacia complanata* (Defrance) is seen throughout the bed. The upper surface of the Anabacia Limestone is generally bored and oyster-covered. The 0.15 m thick Rubbly Beds consist of a biomicrite which locally is crowded with *Sphaeroidothyris sphaeroidalis* (J. de C. Sowerby).

Northwards across the railway viaduct the Upper Trigonia Grit, disturbed by superficial movements, can be seen in several very narrow exposures (7c). Here both behind the old station platform (ST 761 607) and half way along towards the road bridge to the north up to three bands of pelmicrite can be seen interbedded with the silty biomicrudite. The lower fossiliferous band contains many gravel-sized clasts of quartz and fine-grained red sandstone as well as cobble-sized masses of bored, calcareous, sandy siltstone resembling indurated Midford Sands. The whole deposit is extensively bored, but because of the homogeneous texture the borings in the pelmicrite are more easily seen. The 2 m of Upper Trigonia Grit are overlain by the Upper Coral Bed, but this is very poorly exposed.

Proceed along the B 3100, to 50 m west of the Hope and Anchor car park. Here overlying 1 m of Upper Trigonia Grit, 3.5 m of the Upper Coral Bed is intermittently exposed at the roadside (7d). However, the best exposure of the Upper Coral Beds can be seen by proceeding for 200 m up the hill and turning left towards Combe Hay where three metres of the bed is seen 40 m from the crossroads, at ST 757 606 (7e). *Isastrea* colonies make up about 25% of the exposure with the rest of the horizon being nodules of micrite, pelmicrite and oomicrosparite (the latter notably towards the top) set in a cream marly matrix with some very thinly laminated clay bands. Bivalves and gastropods occur throughout the bed.

8 **Tucking Mill** – Smith's Home (Midford Sands) (ST 766 615) Although cars and minibuses can be driven to the NE from Midford towards Monkton Combe, the road is too narrow for coaches. One km along the road Tucking Mill Cottage is seen on the left. This is the house that was accepted by the Bath Natural History and Antiquarian Field Club and the Geological Society of London as the home which William Smith bought in 1798 and owned until he was forced by financial debts to sell in 1819. In fact, the plaque on the wall, described by Cox *et al.* (1941), is on the wrong house. Mrs. Eyles has recently established that the home owned by William Smith was the house 50 m to the NE (Eyles 1974); (see Excursion 13, loc. 9). At the roadside, within 50 m of the correct Tucking Mill House, 4.5 m of Midford Sands is exposed. It is overlain by 0.6 m of disturbed Upper Trigonia Grit. The 'sands' on analysis are very fine sandy silts generally loose or very weakly cemented, but with occasional more cemented bands or 'sand burrs'. Davies (1969) described the Upper Lias sands in southern England but does not describe this particular area.

The disturbed nature of the strata in the Tucking Mill area can be seen by following the signposted footpath NW of Smith's actual home. Although several small exposures show disturbed Midford Sands, after about 40 m is an outcrop showing 2.5 m of Doulting Stone, dipping at about 25° NNE. At 110 m there is 0.7 m of Midford Sands overlain by 2.25 m of Upper Trigonia Grit and 2.2 m of Upper Coral Bed. This area has both slipped and cambered, the dip and fault structures accentuating the dips and giving vertical bed displacements of up to 1.5 m.

In the field at the end of this path is a shallow mudslide (ST 765 618), typical of those on the Fuller's Earth noted by William Smith. It occurred on 11th July 1968 after exceptionally heavy rain in the Bristol-Bath area (Hawkins 1973). The mudslide, on a 12°–14° slope, is 110 m long and has a toe 0.4 m high. The slight depression in which the mudslide took place is frequently wet and thistle-covered. Just beyond the road above the mudslide is a spring, at the base of the Great Oolite.

9 University of Bath (Great Oolite) (ST 767 645) Leaving Tucking Mill it is necessary for those travelling by coach to return to Midford and then take the B 3110 towards Bath. Ascending on to the Great Oolite plateau take the second turning right at the crossroads along Bradford Road towards Combe Down and Claverton Down. After following this road for 5 km and by-passing one entrance to the University, turn right into North Road as you begin to descend the slope. The exposure to be visited is 400 m down this road on the right.

By car from Tucking Mill, continue along the road through Monkton Combe until a crossroads (at ST 780 623). Here turn left up Brass Knocker Hill. Ascending this hill, it is worthy of note that Combe Grove was the site investigated by Smith in 1799, when landslipping damaged much of the estate (Phillips 1844). Indeed, it was here that Smith is first reputed to have recommended the interception of springs by tunnelling into the hillside, reducing the pore pressures and stopping further movement. Having reached the road junction, turn right and after about 2 km, including by-passing one entrance to the University, turn right into North Road. The exposure to be visited is 400 m down this road.

Until 1967 the only outcrop here was the old quarry marked on the 1:25 000 and larger scale maps. A roadway to the University was then cut through the Great Oolite revealing the disturbed state of the 'unfoundered' Great Oolite. Hammers should not be used in the cutting and visitors are asked not to climb on the rock faces, where the University authorities are planting shrubs.

The basal 2 m seen in the northern part of the quarry section are generally medium to thickly-bedded, oobiosparites. Above this is another massive bed of 1.9 m which, especially near the cutting, has weathered to show thickly-bedded strata of oobiosparite grading upwards into oosparite. Poorly seen in the northern cambered and collapsed section of the quarry, but more clearly seen on the left of the cut entrance, is a 0.1–0.2 m band of marl. In the old quarry section, this marl has frequently penetrated up into the disturbed, thinly-bedded 1 m thick overlying pelmicrite bed. Above this is a 0.4 m thick bed of oomicrite with rounded gravel-sized clasts of micrite and having a thin band of pelmicrite. In the quarry section, this extensively bored, more resistant band acts as a type of roof bed. The gulls, often infilled with calcreted limestone fragments, frequently do not penetrate through this bed. Above are 1.25 m of cross-bedded biosparrudites, with the foresets dipping westwards. These cross-bedded rocks are overlain by a 0.35 m bed of biomicrudite which, in the left-hand side of the cutting, is capped by 0.7 m of oosparite.

Of considerable interest in the excavation is the structural disturbance of the beds. Indeed it has not yet been possible to explain satisfactorily all the structures, or to determine accurately the geological succession further than that given above. The gulls, up to 0.4 m wide on the left-hand side of the cutting 15 m from the entrance, are good examples of the way that tension applied to massive beds causes a complete fracture to open; yet in the 2 m thinly-bedded upper horizon, bed-by-bed slip means that the tensional strain is taken up by many small movements and no fracture penetrates through to the

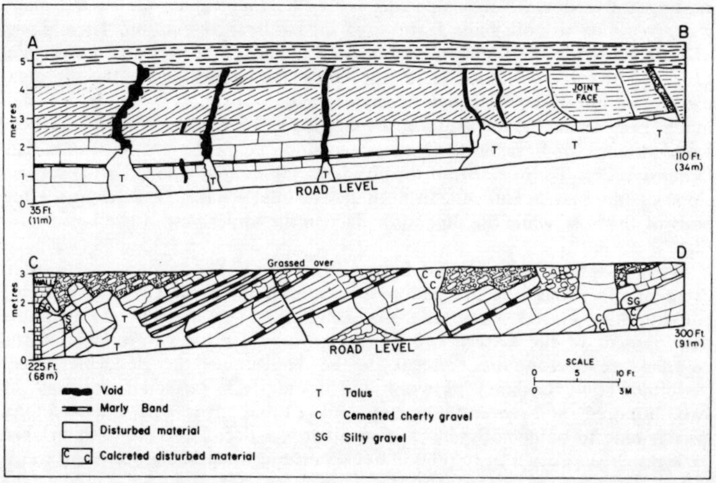

Fig. 3 Section at University of Bath (loc. 9)

surface. The fact that these cavities exist, yet are not visible at the surface, is obviously of importance in the construction of buildings on the plateau surface.

Entering the cut, one of the first things noticed is the almost horizontal bedding to the left, yet the beds on the right have a northerly dip of 26°. This face has not been sampled in detail but attention is drawn to the oosparite bed, over 1 m thick with a highly bored and oyster-covered surface. This distinct bed, when followed laterally is displaced in several places at the position of old gulls, now largely infilled with travertine. Ascending the incline, the sub-horizontal beds on the left-hand side suddenly begin to dip northwards and by the bridge have a dip of 30°. Between the sub-horizontal and inclined beds is a 0.3–0.5 m zone of disturbance, possibly representing a fault or gull breccia. Just east of the bridge, dip and fault structures can be seen displacing by 0.82 m and 1.7 m respectively the 0.43 m bed of biosparrudite with irregular borings overlain by oosparite with fine, generally vertical boring.

Hawkins & Kellaway (1971) drew attention to the infilling deposits in an old cave system and a gull approximately 30 m east of the bridge. Here on the left-hand side of the road a 0.3 m bed of pelmicrite overlies 0.1 m of clayey silt and 0.25 m of gravel in a clayey silt matrix. Although it is impossible to prove, it is suggested these beds are the relicts of a redeposited till of Anglian, or older age. When the cut was fresh it was possible to see that the upper clayey silt was very thinly laminated. It is very improbable that these laminations were deposited at 25° and hence it is suggested that the development of the cave system and the infilling was before the superficial movements which have disturbed the Great Oolite in this area; despite the rocks being shown as unfoundered on the geological maps. Within 2 m of the unindurated gravels are patches of calcareous cemented, highly patinated chert gravels resembling concrete. Twelve metres west of the bridge is a face showing good calcite bands as described by Hollingsworth, Taylor & Kellaway (1944). These are believed to have formed at a time of varying groundwater levels during the Quaternary.

10 North Road (View over Bath) (ST 764 650) Leaving the University cutting travel down North Road. It is worth stopping at one of the many vantage points to note some features of the landscape evolution. The old part of Bath is built in the floor of the valley, on the alluvium overlying Lower Lias Clay. Of importance, however, in the landscape evolution are the slip areas mentioned briefly by Kellaway & Taylor (1968), and now being described in much more detail. To the south of the city is the 46 m high 38°–52° northern wooded slope of Beechen Cliff, the steep slope being the backwall of a Late Devensian slip. To the north of the city is the 34 m high wooded 40° backwall slope of the Beacon Hill slip. In both cases Fuller's Earth Clay forms the top beds of the hills, while the slips took place in the upper part of the Lower Lias Clay.

To the left of the wooded Beacon Hill scar is the asymmetric Camden Crescent. This was not planned to be asymmetric (Ison 1948) but during construction a slip occurred, probably in 1794, which stopped the development of the eastern end of the crescent. Cottages were built on the wooded area below the crescent in the 1860s, but the Hedgemead Slip (redefined by Hawkins) between 1875 and 1885 affected an area of two hundred and seventy houses and caused over one hundred and twenty-nine to be demolished. As a result it was decided to establish an open park rather than attempt to rebuild houses on this disturbed ground. Although other slips can be seen from this stop it is not easy for them to be distinguished.

11 **Beacon Hill** (Inferior Oolite) (ST 751 661) At the bottom of North Road, go 'straight' across into Beckford Road, and at the next road junction turn right into Bathwick Street. At the traffic lights, 150 m past Cleveland Bridge, turn left until the next traffic lights where, by turning right, you begin the ascent to Lansdown. After 400 m turn eastwards into Camden Crescent, where a stop will allow the asymmetric crescent and the wooded área of Hedgemead Park to be observed. At the eastern end of the crescent take St. Stephen's Road to the left and on reaching the hairpin bend, continue straight across to the entrance to The Towers, the head office of M. P. Kent Ltd.

Because of the difficulty of examining the strata on a 30–40° slope, this exposure is only fit for small parties and care should be taken when leaving the footpath to examine the rocks. This previously undescribed site can be examined by climbing the now disused Jacob's Ladder path. The exposure is of interest in that it shows the complete sequence from the Midford Sands to the base of the Fuller's Earth. However, as shown in Fig. 4 considerable bed disturbance has occurred, probably during the Late Devensian landslipping.

The Midford Sands can be seen on the right, a short distance up the footpath. Generally the deposit consists of a weakly cemented very fine sandy silt, with occasional more highly cemented bands – the 'sand burrs'. It will be noted that these show disturbed dips, up to 35°. The junction of the Midford Sands and the Upper Trigonia Grit can be seen just to the left of Jacob's Ladder. Here 0.4 m of Midford Sands is exposed. Within the top 0.2 m there are some cemented nodules which merge upwards to form a layer of calcareous siltstone. The Upper Trigonia Grit has a tripartite sub-division. The lower band is a silty biomicrudite with indications of considerable bioturbation. This band contains pebbles of quartz and fine-grained red sandstone. Several masses of calcareous siltstone similar to that at the top of the Midford Sands exist in the lower part. The position of the bored surfaces indicates that some of these are over-turned. The rock is frequently cavernous due to the solution of shells, but terebratulids, gastropods, crinoid ossicles, *Trigonia* sp., and part of the ammonite, *Garantiana garantiana* (d'Orbigny) occurred within 0.12 m of the base. At the top of the bed, some of the larger bivalves, notably *Trichites* sp., have been bored by annelids and lithophagi.

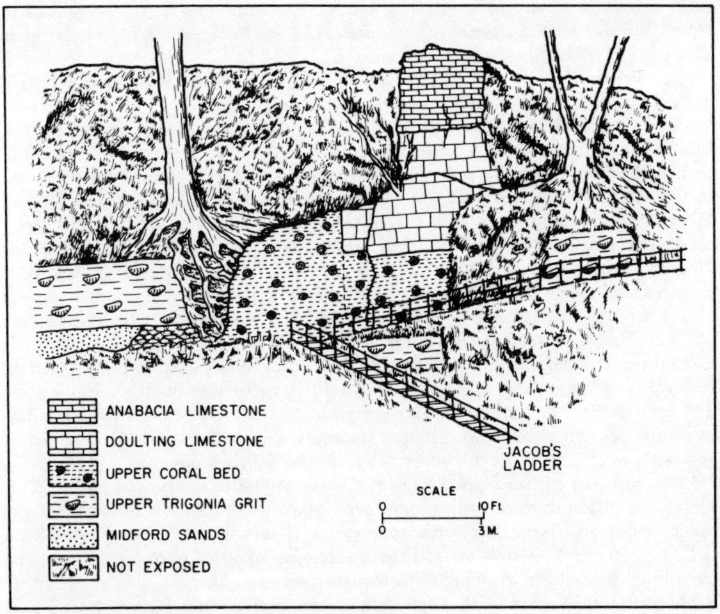

Fig. 4 Details of geology at Beacon Hill (loc. 11)

The middle band of the Upper Trigonia Grit is a silty pelmicrite. The upper and lower junctions are poorly defined, but differential weathering has made two slightly irregular bedding planes more prominent. Macrofossils are rare but the whole bed is extensively bored. The top bed is a slightly ferruginous biomicrudite. Phyla similar to the lower bed occur and the ammonite *Heptosphinctes meseres* (S. Buckman) was recovered 0.2 m from the base. The top part of the bed is extensively bored.

The Upper Coral Bed is best developed to the east, but at the same height as the outcrop just described. Here a small slip, 13 m across, has displaced the beds by about 3 m. About 2 m of the Upper Coral Bed is exposed. The appearance is dominated by the harder crystalline masses of *Isastrea* sp., 0.2–0.5 m long and up to 0.1 m high, lying almost horizontal and concave upwards. Coral occupies about a fifth of the rock volume, the rest being composed chiefly of nodules of micrite and biomicrite set in a marl. Sometimes near the top lumps of oosparite occur. Bivalves, gastropods, and polyzoans can be seen.

The base of the Doulting Stone is clearly defined. The stone has a very massive appearance and begins with a biopelsparite with serpulids, passing upwards into an oosparite with pisoliths. About 1 m from the top, the bed has a nobbly appearance due to the presence of large intraclasts, while in places the hollow moulds of bivalves accentuate the uneven surface texture.

The Anabacia Limestone begins with an oosparite with pisoliths and some intraclasts up to 5 cm across. The lower 0.4 m of the pelmicrite is harder in texture and has been bored. The upper pelmicrite contains some *Anabacia complanata* (Defrance) while occasional moulds of *Trigonia* sp. and serpulid borings occur.

The Anabacia Limestone and Rubbly Beds can be seen both at the top of this exposure and also to the left slightly further up the footpath. The Rubbly

Bed is only 0.15 m thick. It is a biomicrite with abundant terebratulids, especially *Sphaeroidothyris sphaerodalis* (J. de C. Sowerby). Overlying is about 0.5 m of Lower Fuller's Earth Clay.

12 **Brown's Folly** (Great Oolite) (ST 795 661) From Beacon Hill return back to the A 4 road and proceed out through Bath in the direction of London. At Bathford leave the A 4, pass under the railway bridge on the Bradford-on-Avon road (A 363) and turn left towards Kingsdown. Near the end of the houses on the right-hand side there is a steep road (Prospect Place) to the south. Coaches should not attempt this road, but cars may proceed the 400 m to a parking place (ST 798 664) on the south side of the very steep road. The path to the Folly is fairly clearly defined and should be followed. This exposure has recently been referred to by Green & Donovan (1969) who record a succession thicker than that easily seen at present.

At the top is about 4.7 m of current-bedded biomicrudite representing part of the Forest Marble. The rock is often cavernous and locally bored. Sometimes it becomes an interbedded (2–35 mm thick) oomicrite/biomicrite. Beneath a poorly exposed area is another 1.5 m of dominantly biomicrites. The marls, micrites and biomicrites representing the level of the Bradford Clay can only be seen beneath overhangs. However, *Digonella digona* (J. Sowerby), and parts of *Apiocrinus* etc. can be found in the debris below.

The top bed of the Upper Rags of the Great Oolite is a biosparrudite. The surface is often bored and oyster-covered while the rock contains corals, brachiopods and large thick-shelled bivalves (Green & Donovan 1969). This bed is channelled up to 1 m into the underlying band of oomicrosparite with occasional intraclasts, in which macrofossils are rare. The massive 'roof bed' is a harder biomicrudite with some microspar cement and with a bored and oyster-covered upper surface. Below the 'roof bed' is 2 m of oomicrosparite, representing the Bath Stone. The rest of the sequence given in Green & Donovan (1969) is not readily visible.

13 **A 46 Swainswick Road Cutting** (Fuller's Earth) (ST 756 691) From Brown's Folly return towards Bath along the A 4 and at the traffic lights turn right, up the A 46 road at Lambridge. After 2.5 km, just north of the village of Swainswick, the road enters a cutting constructed in the early 1960s, due to the old road continually slipping down the hillside. The transport should be parked along a section of the old road on the left, at the top of the cutting.

The 35° slopes of the cutting are now generally grassed over, but because of a number of slips which occurred in the heavy July rainstorms of 1968, it is possible to see the weathered Fuller's Earth Clay and to collect fossils. The main fossil horizon exists about half-way down the cutting where *Liostrea acuminata* (J. Sowerby) occurs in great profusion and such brachiopods as *Rhynchonelloidella wattenensis* Muir Wood and *Sphaeroidothyris sphaeroidalis* (J. de C. Sowerby) occur quite commonly.

Half-way up the cutting the excavating contractor tipped material in the valley to the west. During the heavy July 1968 rainstorm this slid and flowed down the hillside. From this vantage point it is possible to look across the Swainswick Valley and see the large crescentic slope of Soper's Wood. In the field to the south of the wood is another slip, activated in 1968. Nearby, on slopes of 12–16°, turf rolling and surface creep have occurred during the wet spells of most recent winters. With care it is possible to see gaps in the trees at the top of the wood; here mudslides in the Fuller's Earth are still active and during periods of wet weather transport trees up to 0.25 m in diameter.

13
William Smith localities

R. Bradshaw

Objectives

To visit geological localities, works and buildings in the Bath area associated with William Smith.

Special Features

The houses mentioned in the itinerary are all in private hands and are not open to viewing. Several of the roads followed on the excursion are narrow and winding and not suitable for coaches.

Maps

Ordnance Survey	One-Inch Sheet 166 (Frome)
	1:50 000 Sheet 182 (Bristol & Bath)
	1:25 000 Sheets ST 65, ST 75, ST 76, ST 85
Geological Survey	One-Inch Sheets 265 (Bath), 281 (Frome)

Main Reference

Cox (1941).

Introduction

William Smith is widely regarded as the 'Father of English Geology'; his friends and pupils called the Bath area the 'Cradle of Geology'; and he himself called Rugborne Farm (see later) the 'Birthplace of Geology'.

At the age of twenty-two Smith came to Somerset in October 1791 to begin a ground survey of an estate near High Littleton and underground surveys at Mearns Colliery (Phillips 1844, pp. 6–7). In March 1793, he was requested to prepare preliminary levels for a proposed canal to link the Somerset coalfield with the Kennet and Avon Canal and thus with London. The canal was to have two branches (Fig. 1), a northerly one beginning at Paulton and a southerly one beginning at Radstock, to join the other at Midford, and then join the Kennet and Avon north of Limpley Stoke.

The map of the route was completed in November 1793 and in the following year the Canal Bill was passed through Parliament and Smith was appointed engineer and surveyor to the Company. The troubled history of the construction of the canal from 1795–1805, its operation and its closing to commercial traffic in 1893 because of competition from the railways, can be found in accounts by Clew (1970) and Buchanan & Cossons (1969) while Smith's work in Somerset is described by Eyles (1969) and Cox (1942).

Outline Geology

The geology of the area is illustrated in Fig. 1 and a diagrammatic section in Fig. 2. As early as 1714 John Strachey had recognised the pronounced unconformity between the steeply dipping and often highly deformed Carboniferous strata and the Mesozoic rocks with their gentle dips to the east. Smith too appreciated this structure and the fact that as the canal was driven eastwards newer rocks with different fossil assemblages were encountered and that they had a regular dip. It was here then that he first realised the significance of 'organised fossils'.

The stratigraphic table drawn up by Smith in 1799 and subsequent modifications in 1812 and 1815–16 are illustrated by Phillips (1844, p. 146). The modern version of the succession of rocks in the vicinity of the canal is given in Table 1.

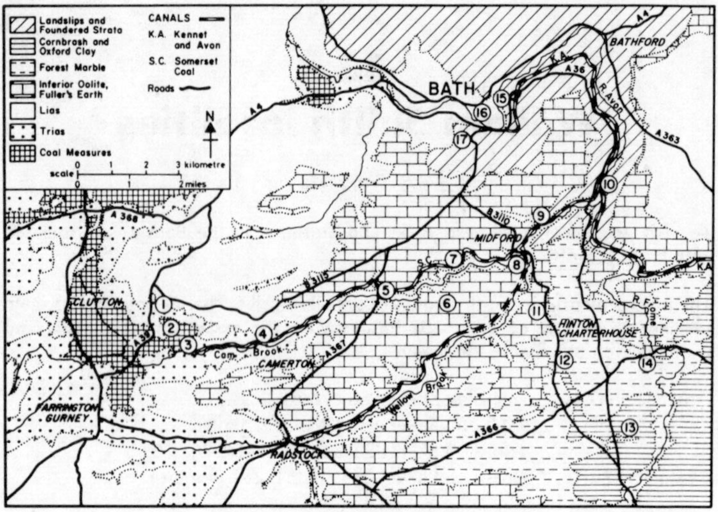

Fig. 1 Simplified geological map showing the line of the Somerset Coal Canal and the proposed line of the southern branch.

Itinerary

Take the A 37 Shepton Mallet road southwards from Bristol through Pensford with its old coal tips and impressive railway viaduct, Temple Cloud with abandoned quarries in Pennant Sandstone, to the junction with the A 39 at White Cross. Turn left and proceed through High Littleton to Cross Ways where turn right along the B 3115 and after about 1 km turn right on to a narrow minor road.

1 Mearns (ST 652 587) This is the site of perhaps the earliest colliery surveyed by Smith, in 1792 (Bulley 1952; Phillips 1844). Houses have been built on the site of the colliery but the old flattened tips exist to the east.

Continue along the minor road to its junction with the A 39, turn left and after 350 m, just before the church turn left again on to a minor road signposted Timsbury and Camerton. After another 350 m towards Timsbury there is a signpost in the hedge stating 'William Smith lived here'.

2 Rugborne House formerly Rugborne Manor (ST 651 583) In this seventeenth century house built on the Upper Coal Series with nearby Trias and Lias, Smith lodged from 1792–5, the years in which he first recognised the fundamental truths of stratigraphy and where he determined the succession of strata. This is the 'Birthplace of Geology'. In a manuscript diary quoted by Cox (1941) Smith recorded that he 'resided in part of the large old manor house belonging to Lady Jones, called Rugburn. It was then occupied by a farmer who lodged and boarded me for half a guinea a week and kept my horse for half a crown a week'.

Continue along the minor road towards Timsbury and take the first turning right and first left to Goosard Bridge and the entrance to a refuse tip and sewage works belonging to Clutton Rural District Council (Fig. 3a).

3 Paulton Colliery (disused) (ST 654 576) Walk along the road with the sewage works on the right and then just before the old tip turn left towards the valley of the Cam Brook. Immediately in front is the partially filled-in terminal basin of the Somerset Coal Canal with a leat running from a brook, a ruined cottage which was occupied by the 'lock keeper', and the loading basins to which tramways ran from Mearns, Hayeswood and Paulton Collieries. Here

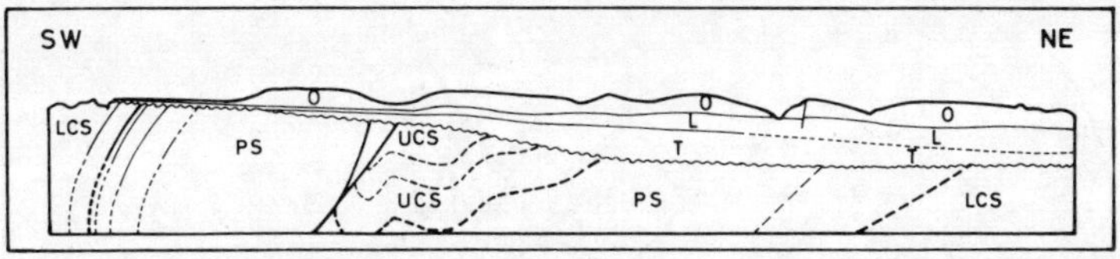

Fig. 2 Generalised SW–NE cross section from south of Radstock to near Midford. LCS – Lower Coal Series, PS – Pennant Sandstone, UCS – Upper Coal Series, T – Trias, L – Lias, O – Oolites.

			Thickness in metres
		Oxford Clay	up to 160
		Kellaways Clay	up to 27
	Great Oolite	Cornbrash	3 – 6
		Forest Marble	21 – 40
		Great Oolite Limestone	0 – 34
		Fuller's Earth	35
JURASSIC		Inferior Oolite	9 – 18
	Lias	Midford Sands	0 – 30
		Clays	up to 46
		White and Blue Lias	0 – 15
TRIAS	Keuper	Rhaetic	3 – 9
		Tea Green and Red Marl	up to 60
		Dolomitic Conglomerate	up to 24
	UNCONFORMITY		
CARBONIFEROUS	Coal Measures	Upper Coal Series	210
		Pennant Series	210
		Lower Coal Series	120
	Millstone Grit	Quartzitic Sandstone	6
	Limestone		180

Table 1

too is the site of Paulton Foundry, the only major foundry in the coalfield (Down & Warrington 1971, pp. 87–91).

Take the minor road southwards from Goosard Bridge and on the outskirts of Paulton turn sharp left towards Radford and Camerton. Park near the Jolly Collier Inn.

4 **Camerton** (ST 679 579) Walk over the bridge over the old railway and see the site of the canal in the fields to the west (a footpath runs along the bank) and in front of the Post Office. Fig. 3b shows the layout at the end of the nineteenth century of the two Camerton pits. Little is now left of the collieries but the canal is visible and there is a good view from the top of the New Pit tip (ST 687 582) which can be reached by a minor road near the chapel (Down & Warrington 1971, pp. 112–7).

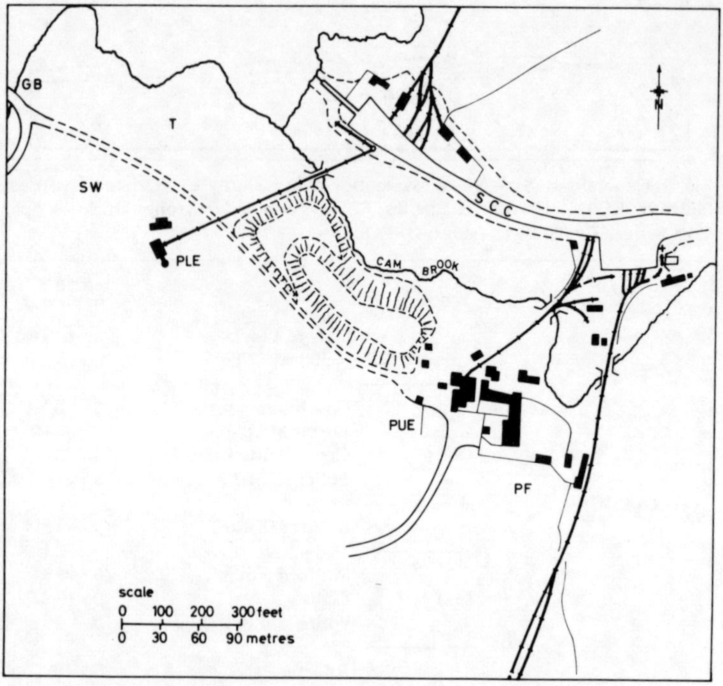

Fig. 3a Paulton collieries and the western end of the Somerset Coal Canal (about 1860). SCC – Somerset Coal Canal, PLE – Paulton Lower Engine, PUE – Paulton Upper Engine, PF – Paulton Foundry, SW – modern sewage works, T – modern tip, GB – Goosard Bridge.

Continue eastwards along the valley of the Cam Brook through Carlingcott to Dunkerton with traces of both the canal and old railway visible.

5 **Dunkerton** Turn left on to the A 367 and park in the big lay-by 250 m from the junction (ST 717 594). Just below the northern end of the lay-by is a farm which is built right across the canal which can be seen swinging round the small valley to the west and passing beneath the main road where there is a parapet of an original bridge. Walk southwards along the main road and on the left find Swan Villa, formerly the Swan Inn where Smith stayed several times. He wrote his earliest surviving notes on fossils, and dictated his *Memorandum on Fossils* in January 1796 (Phillips 1844, pp. 17–22).

Drive northwards along the A 367 and after about 1 km turn right at the Cross Ways Inn towards Combe Hay then at the crossroads just over the old railway turn right towards Wellow then left on the top of the hill towards Twinhoe.

6 **Overgrown Quarry** (ST 740 592) in the Great Oolite which was visited and described by Smith (Cox 1941, pp. 24–5) (see Excursion 12, loc. 4).

7 **Combe Hay Locks** (ST 738 603 – ST 748 604) Return to the crossroads and into Combe Hay, take the road to Midford. Continue about 1 km beyond Combe Hay and park by the old railway bridge opposite a blue house (ST 7455 6040). From here a path leads into the woods (Fig. 4) and alongside the chain of locks. This path can be followed to emerge on the road again by Caisson House. If you wish to explore the system fully beyond the

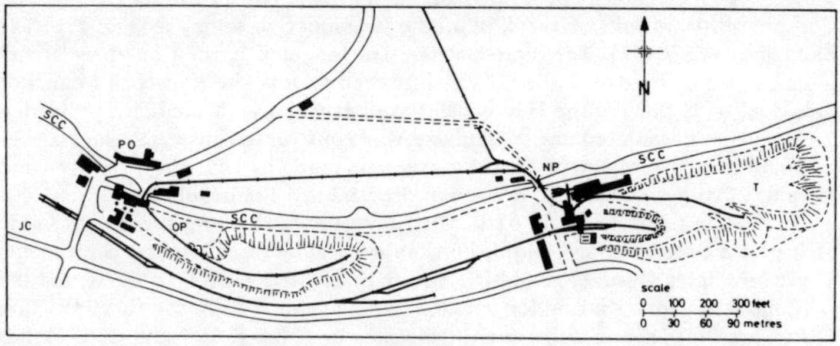

Fig. 3b Camerton collieries in 1883. OP – Old Pit, NP – New Pit, SCC – Somerset Coal Canal, T – Tip, PO – present Post Office, JC – Jolly Collier Inn. (After Down & Warrington 1971)

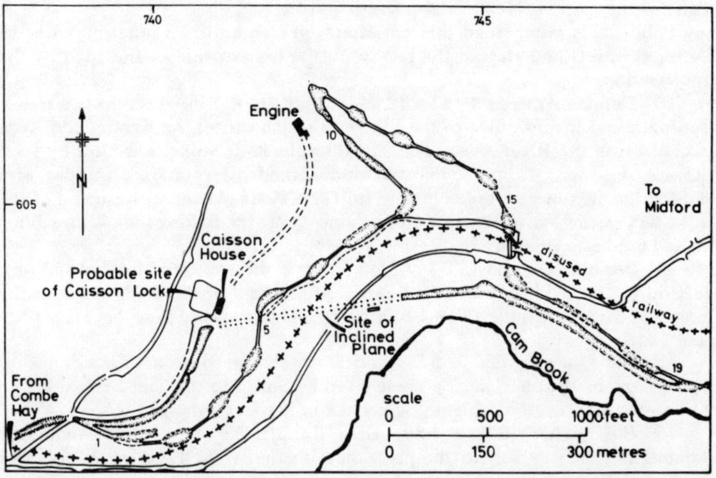

Fig. 4 Locks, inclined plane and probable site of Caisson Lock near Combe Hay (after Buchanan & Cossons 1969). Dashed line is the Limpley Stoke Railway (1910).

paths, obtain permission from Mr. McArthur at Caisson House.

The drop in level of the canal between Paulton and Limpley Stoke is about 40 m and it was proposed that the whole of this drop should be concentrated near Combe Hay (Buchanan & Cossons 1969, pp. 190–2; Clew 1970, pp. 23–64 & pp. 160–9). Details of the three methods employed must be sought in the references cited. The sites of the Caisson Lock and part of the inclined plane are not now clear but the flight of twenty-two locks, which was the ultimate solution to the problem, is well preserved. There is a good section in the Inferior Oolite along the disused railway line. Here, as at several other places, the railway uses part of the bed of the canal. It was begun in 1907 opened in 1910 and closed in 1951 but was temporarily revived in 1952 as the setting for the film 'The Titfield Thunderbolt'.

8 Midford Aqueduct (ST 7575 6050) Continue along the road towards Midford, turn right on to the B 3110 and almost immediately right again. First cross the metal bridge over the Somerset Coal Canal (now a garden here), then the stone bridge over the Cam Brook. Just before the dismantled railway bridge over the road, a track leads off to the right (west) through a field and alongside the brook. After 250 m the aqueduct is seen (see Fig. 2, p. 126 Excursion 12, loc. 7). The aqueduct is 20 m long and carried on three arches inscribed with the date 1803. It was intended to link the Radstock branch of the canal with the Combe Hay canal, but difficulties with the locks prevented it ever being completed and a tramway was constructed instead. The transfer of coal from the tramway to the canal was made here at the aqueduct and remains of spilt coal can still be seen in the banks of the brook.

9 **Tucking Mill** (ST 765 616) Then take the road along the northernside of the Cam Brook towards Monkton Combe to Tucking Mill where Smith lived at various times from 1798–1819. The Bath Natural History Society and the Geological Society of London erected a plaque on the house in 1932 (Cox 1941, pp. 26–7) but subsequent research by Mrs. J. M. Eyles (1974) has shown that the house owned by Smith is about 100 m to the east. Near to this house is a road section in Midford Sands from which he collected specimens and on the south side of the road there is a bridge over the canal with just a little of the span visible. (See also Excursion 12, loc. 8).

Follow this minor road through Monkton Combe to its junction with the A 36, turn left and stop in the lay-by on the eastern side of the road by the intersection.

10 **Dundas Aqueduct** (ST 792 644) Opposite the intersection is a public footpath which runs down to the aqueduct which carries the Kennet and Avon Canal across the River Avon. The aqueduct, in Bath Stone, was built by John Rennie, engineer to the company, and named after Charles Dundas who guided the fortunes of the company for forty years. At the western end of the aqueduct is the basin where the two canals join; the linking lock is now filled in and used as a rose garden.

11 **Broadfield Farm** (ST 767 589) Retrace the route to Midford and turn left on to the B 3110. Broadfield Farm, on the right of the road, was once the home of Smith's brother, John, and he frequently stayed here between 1805 and 1819.

12 **Old Quarry** (ST 772 573) Continue through Hinton Charterhouse to the quarry in Hinton Sands, a name given by Smith to the beds here which he put above the Forest Marble but which are in fact equivalent to it.

13 **Old Quarry** (ST 794 550) Then through Norton St. Philip with the famous old George Inn to the junction with the A 36. Take the minor road leading NE toward Tellisford and on the left is a quarry with Lower Cornbrash, Forest Marble Clay and Forest Marble from which Smith named a number of Cornbrash fossils.

14 **Farleigh Hungerford** (ST 800 565) Here is the rectory which used to be the home of Smith's friend, the Revd. Benjamin Richardson.

15 **29 Pulteney Street** From Farleigh rejoin the A 36 to Bath along the Warminster Road, round Sydney Gardens to 29 Pulteney Street on which there is a plaque to commemorate the dictation of the 'Table of the Order of Strata and their embedded Organic Remains in the vicinity of Bath, examined and proved prior to 1799'. The Revd. J. Townsend, owner of the house, and the Revd. B. Richardson were the two friends who had the privilege of noting down this early stratigraphic table (Phillips 1844, pp. 29–31).

16 **Trim Bridge** Along Pulteney Street and Argyle Street is Upper Borough Walls and Trim Bridge. Smith had his office here from 1802–5; No. 3 has been considered his office but Dr. H. Torrens (pers. comm.) has discovered evidence that the house used by Smith was No. 2 Trim Bridge. Here there was a collection of fossils unique in that they were stratigraphically arranged and displayed in boxes on the floor for the instruction of the many famous visitors; the collection is now on display at the British Museum (Natural History).

17 Bloomfield Crescent The Wells Road A 367 leads up the south side of the Avon Valley. After two major bends it is joined from the right by Bloomfield Road off which on the right lies Bloomfield Crescent. The middle of the three houses was Smith's home from 1795–1798. From this house one can look across the Avon valley and see the sites of some of the major landslips for which Bath is famous and which are described in Excursion 12. High rainfall in 1799 led to a number of such slips and Smith carried out effective remedial measures at Batheaston, Combe Grove and a number of other places (Phillips 1844, pp. 32–4). He also restored the Hot Springs in 1810.

14
The south and mid Cotswolds

J. W. Murray and P. L. Hancock

Objectives To examine the Mid Jurassic stratigraphy of the Cotswold Hills, with emphasis on the facies variations of the Inferior Oolite. Some localities also display Lias and Great Oolite, and in the area there is a good range of limestone types to be seen. The palaeontology and structural geology are also included.

Special Features Cars can be taken close to all localities, but the approach roads to localities 2, 14, 15, 18 and 19 are unsuitable for coaches, though all are within easy walking distance of the main roads. Locality 4 can only be approached by coach along the road off the A 4135 from the east.

Permission should be obtained in advance to visit the following:

Breakheart Hill Quarry: T. F. Coke Ltd., 267 Gloucester Road, Cheltenham. Cheltenham 2774.

Pyke Quarry: Stroud Quarries Ltd., Horsley. Uley 421.

Catbrain Quarry: Painswick 2131.

Maps

Ordnance Survey	One-Inch Sheets	143 (Gloucester & Malvern)
		156 (Bristol & Stroud)
	1:50 000 Sheets	162 (Gloucester & Forest of Dean)
		163 (Cheltenham & Cirencester)
	1:25 000 Sheets	SO 80, SO 81, SO 91, ST 79, ST 89
Geological Survey	One-Inch Sheet	251 (Malmesbury)
	1:50 000 Sheet	234 (Gloucester)

Outline Geology The Cotswold Hills extend from near Bath in the south to near Chipping Campden in the north. They represent the escarpment of a series of limestones of Middle Jurassic age which overlie the softer sands and clays of the Upper Lias. The highest point of the escarpment is at Cleeve Hill near Cheltenham (326 m). To the south the general height decreases to an average of 183–229 m near Bath. In the southern part, the escarpment is fairly linear but north of Hawkesbury it is deeply eroded to form the steep sided valleys known as 'coombes'.

The earliest geological reference to the Cotswolds is that of Conybeare & Phillips (1822). Richardson (1910) has given a comprehensive description of the history of investigation of the Cotswold Inferior Oolite. During the nineteenth century numerous workers were describing sections and attempting stratigraphical correlations. Amongst these may be mentioned Murchison, Strickland, Lycett, Wright, Hull, Witchell, J. Buckman, S. S. Buckman, Hudleston and Woodward. However, it was through the detailed studies of the ammonite faunas that S. S. Buckman was able to make the greatest contribution to the understanding of the Upper Lias – Inferior Oolite stratigraphy (monographs 1887–1907; stratigraphy 1895, 1897, and 1901). Where ammonites were rare Buckman made use of the abundant brachiopod faunas (1895, pp. 439–461). His work was extended by Richardson (1910). Since that time no major study of the Inferior Oolite has been undertaken although there have been short reports of new or undescribed exposures (Channon 1950; Murray 1969) and numerous reports of field excursions to the area by the Cotteswolds Naturalists Field Club and the Geologists' Association (see McKerrow *et al.* 1964; Ager *et al.* 1973). In addition Arkell (1933) provided a good summary of the Cotswold stratigraphy and fitted it into the regional framework of Britain.

Few of the Cotswold quarries described in this book are still being worked; most had ceased to be worked by the beginning of the century (Richardson 1910). Due to the normal processes of weathering many of the sections have deteriorated and become overgrown to a greater or lesser extent. Further, the over-enthusiastic hammering activities of geological parties has led to the build-up of extensive screes at the foot of quarry faces and these tend to obscure more and more of the section (as for instance in the Nibley Knoll Quarry).

Before describing the sections that can be visited, it is necessary to comment on the stratigraphy. The itineraries proposed here are concerned primarily with exposures in the Inferior Oolite but locally there are quarries in the underlying Middle and Upper Lias and in the overlying Great Oolite which have been included too.

The Middle Lias is represented by the Marlstone, an impure ferruginous limestone, which gives rise to the lowermost escarpment and terrace at the foot of the main Cotswold scarp around Wotton-under-Edge, Stinchcombe, Dursley and Cam Long Down. Above follows some 12 m of Upper Lias clays which are never seen in outcrop. Then follows the Cotteswold Sands – silty, micaceous, fine-grained sands, for the most part soft but locally hardened into irregular rock bands particularly in the lower part. These sands form part of a diachronous series of sands running south (Midford Sands, Yeovil Sands) to the Dorset Coast (Bridport Sands). This classic example of diachronism was established through the detailed studies made by Buckman between 1875 and 1922. He showed that the sands were oldest in the Sodbury to Stroud area and that they became younger in passing south. Bridport Sands were shown to be the same age as the Cephalopod Bed, a condensed series of ferruginous marls and limestones restricted to the southern Cotswold area and succeeding the Cotteswold Sands there. In the north Cotswolds the deposits of this age are clays – Upper Lias Clay.

The succeeding Inferior Oolite is a series of limestones of Aalenian and Bajocian ages. The standard succession and bed names for the Cotswolds are shown in Table 1.

The majority of earlier lithological descriptions are either misleading or uninformative. Grit was used to denote coarser grained limestones, ragstones are richly fossiliferous limestones, and most of the freestones are oolitic. The range of rock-types represented is very great and in preparing these descriptions every rock band in each of the localities discussed has been sampled, polished, etched, stained and studied from acetate peels. All the limestones are defined according to the classification described by Folk (1959, 1962), the details of which are explained below. This, together with the known distribution of macrofossils, aids the interpretation of the conditions of deposition of the beds.

It has been found to be necessary to consider the Inferior Oolite on a detailed bed-by-bed basis because it is not otherwise possible to understand the stratigraphy very readily.

The Scissum Beds are everywhere sandy biomicrites and they rest either on similar deposits (the upper part of the Cephalopod Bed in the region from Wotton-under-Edge to Frocester Hill) or on Upper Lias clay as at Crickley Hill and Leckhampton Hill. In each case the marked increase in the carbonate content probably indicates a decrease in the supply of clastic material rather than a sudden increase in biological production. The abundant bivalve and echinoderm (particularly crinoid) bioclasts indicate a shallow shelf

STAGE	ZONE	FORMATION	LITHOLOGY	GROUP
BAJOCIAN	Parkinsonia schloenbachi	*Clypeus* Grit	intramicrite, biomicrite, oosparite	Upper Inferior Oolite
	Strigoceras truelli	non-sequence		
	Garantiana garantiana	Upper *Trigonia* Grit	biomicrite	
	Strenoceras niortensis			
	Teloceras blagdeni	unconformity		
	Otoites sauzei	*Phillipsiana* Beds *Bourguetia* Beds		Middle Inferior Oolite
	Witchellia spp.	*Witchellia* Grit Notgrove Freestone	oomicrite	
	Shirbuirnia spp.	Gryphite Grit	sandy biomicrite	
	Hyperlioceras discites	*Buckmani* Grit Lower *Trigonia* Grit	sands and biomicrite biomicrite	
UPPER AALENIAN		unconformity		Lower Inferior Oolite
	Ludwigella concava	Tilestone Snowhill Clay Harford Sands		
	Brasilia bradfordensis	Upper Freestone Oolite Marl	oosparite variable	
	Ludwigia murchisonae	Lower Freestone Pea Grit	oosparite pisolith biomicrite	
	Ancolioceras spp.	Lower Limestones	oosparite	
	Tmetoceras scissum	*Scissum* Beds	sandy biomicrite	

Table 1 Details of the Inferior Oolite Series generalised for the whole Cotswolds

environment.

The Lower Limestone in the Wotton to Selsley Common area is an oosparite, often massive, and sometimes showing both large and small scale cross-bedding. It can be interpreted as forming in a shoal, high energy, subtropical or tropical, marine or slightly hypersaline, environment. The ooliths probably formed submarine banks (as indicated by the large scale cross-bedding) upon which there were smaller ripples and mega-ripples (represented by the small scale cross-bedding). In the Crickley Hill – Leckhampton Hill area the Lower Limestone is thinner and is represented by biomicrites and biomicrudites indicating open shelf conditions.

The Pea Grit Series is best developed around Crickley Hill where numerous massive beds of pisolith-bearing biomicrites are to be seen. These beds have long been famous for their rich fauna particularly of brachiopods and echinoids. The pisoliths may be 1 cm or greater in diameter. Many are flattened ellipsoids with a core of bivalve or brachiopod shell fragment while others are nearer a spherical form. Their growth is attributed to the activities of the encrusting calcareous alga *Girvanella pisolitica* (Wethered 1891) and the fact that the coatings completely enclose the core indicates that the pisoliths were periodically rolled over (perhaps during storms). By analogy with modern regions of carbonate sedimentation (e.g. the Trucial coast of Arabia) the presence of echinoids would indicate an open marine environment or close proximity to open sea (as in a bay or at the entrance to a lagoon). Arkell (1933, p. 204) considered the environment to have been a coral reef but corals are not the most conspicuous components of the rock. It seems probable that the environment was just sublittoral with perhaps a metre or so of water, and that it was too disturbed to allow the development of a continuous algal cover to form a bedded stromatolite, but not so disturbed as to prevent the growth of the algae on the bioclasts. This area was probably an embayment, for to the north and south the pisoliths die out. In the Wotton-Nibley area a Pea Grit 'equivalent', usually less than 30 cm thick, has been recognised but it seems likely that part of the Lower Limestone here is equivalent in age to the Pea Grit of the Crickley area.

The Lower Freestone is typically an oosparite developed on a massive scale particularly at Leckhampton where sedimentary structures are very well shown. Its environment of formation must have closely resembled that of the oosparitic Lower Limestone.

The Oolite Marl consists of an alternation of soft marls and thicker hard bands of limèstone. In its southermost occurrence (Selsley Common) these are biomicrites, with or without ooliths, and oosparites or biosparites sometimes with intraclasts and grapestones. At Leckhampton they are micrites, pelmicrites and oomicrites. On the whole a low energy, shoal lagoonal environment is indicated.

The Upper Freestone is an oomicrite or oobiosparite at Selsley Common and an oosparite at Leckhampton. The environment of deposition was probably similar to that of the other oolites.

No other beds of Lower Inferior Oolite age are exposed in the area under discussion, although they occur in Cleeve Hill to the north. Their absence from the south may be attributed in part to the effects of slight folding followed by erosion and in part due to non-deposition (particularly in the south, see Buckman 1901). Two erosional phases are recognised; one at the base of the Bajocian and one in the Middle Bajocian (formerly known as the Vesulian transgression). South of Stroud there are no Lower Bajocian deposits (i.e. Middle Inferior Oolite) and the erosion surface is attributed to the Middle Bajocian.

Rocks of Middle Inferior Oolite age are seen at Catbrain Quarry and on Leckhampton Hill. The Lower Trigonia Grit consists of ferruginous biomicrites sometimes with chamosite-oolith bands. The Buckmani Grit cannot be recognised at Leckhampton but is represented by sands and sandy biomicrites at Catbrain Quarry. The Gryphite Grit is a sandy biomicrite rich in *Gryphaea sublobata.* The Notgrove Freestone is an oomicrite. All these formations are thin in the area described here.

Following the Middle and preceeding the Upper Inferior Oolite depositional phases there was slight folding accompanied by erosion. This led to the Middle Bajocian unconformity (= Vesulian transgression of some authors) upon which the Upper Trigonia Grit rests. Buckman (1901), in a classic piece of mapping and geological deduction, demonstrated that the pre-Upper Inferior Oolite rocks were thrown into a series of very gentle fold-like structures with axes varying from NW–SE (as in the Painswick 'syncline') to more or less N–S.

The Upper Inferior Oolite commences with the Upper Trigonia Grit, typically a richly fossiliferous biomicrite (*Acanthothyris,* terebratulids, and *Trigonia*) indicating open shelf conditions. However, at Breakheart Hill there are oobiosparites suggestive of shoal high energy conditions, while at Broadway, Dursley, there is a thicker succession of intramicrites indicating a nearby source of reworked intraclasts. A bored surface near the top of the formation in Leigh's Quarry, Selsley Common, indicates a local brief interruption of sedimentation.

The interruption may be followed by a bed recognised as the Upper Coral Bed by analogy with the succession at Dundry (as in Richardson 1910). The only localities where this bed could be distinguished are Nibley Knoll and Wotton Hill. However, it is doubtful whether this can truly be correlated with the Dundry succession.

The highest formation is the Clypeus Grit. (Reference to Table 1 shows a non-sequence with the Upper Trigonia Grit.) To the south, at Coombe Hill, Wotton, Breakheart Hill and at Broadway, Dursley, it consists of intramicrites, biomicrites and oosparites. The stiphrothyrids and *Acanthothyris* seem to be associated with the intramicrites suggesting that they lived attached to the rocks which were being eroded to provide the intraclasts. Their good preservation suggests only limited transport from the life habitat. The succession is capped with oomicrite or oosparite. To the north, at Stanley Wood, biomicrites appear. On Selsley Common biomicrites and biosparites predominate but intraclasts appear throughout the succession and testify to the erosion of nearby limestones. The Clypeus Grit gives evidence of shoaling conditions at the conclusion of Inferior Oolite time.

The Bathonian commences with the Acuminata clays, named after the common occurrence of *Liostrea acuminata*. No exposures are available at present. Conditions soon returned to carbonate sedimentation leading to the formation of the 'Great Oolite'. This has been subdivided into a number of units by Arkell & Donovan (1952) and some of these may be seen at the Pike Quarry, Horsley, and Burleigh near Minchinhampton.

The structural geology of the Cotswolds is discussed in detail in Hancock (1969). This summary is based on that account. The gentle E or SE tilt of the Jurassic limestones is complicated by six N–S or NW–SE trending gentle folds of intra- or post-Jurassic age. Four sets of normal faults belonging to two systems displace the limestones. Two conjugate sets of faults strike 10° north of west while the other two conjugate sets strike N–S.

The principal types of minor tectonic fractures are joints, calcite veins and lineated surfaces. The last category includes surfaces bearing nearly horizontal slickenside or slickolite lineations. Slickenside lineations are developed on rock surfaces and in calcite veins. The lineation formed by slickolites generally penetrates up to 2 cm into the limestone adjacent to the surface. Both types of lineated surface also exhibit small asymmetric steps perpendicular to the lineation. Both normal and reverse steps are represented, they allow the sense of shear along surfaces to be determined.

In horizontal limestones the complete minor fracture system comprises six sets, of which four are vertical and two are moderately inclined. The whole system is symmetrically arranged about the regional N–S strike of the beds. Two of the vertical sets or a conjugate system of horizontally lineated wrench shear surfaces, striking NE (set J_1) or NW (set J_2), and enclosing an average dihedral angle of 82° about an E–W bisectrix. Their development resulted in a slight E–W shortening and a slight N–S elongation of the thin limestone prism of the Cotswolds. The other two sets of vertical fractures are orthogonal extension joints and veins striking N–S (set J_3) and E–W (set J_4). The two moderately inclined conjugate sets of normal shear joints are restricted to the neighbourhood of Painswick. Both sets strike E–W, with one set dipping north at about 55° (set J_5) and the other dipping south at about 55° (set J_6). Although the trends of sets J_1 to J_6 commonly follow these directions, at some localities the whole system is arranged up to 25° clockwise or anticlockwise of the usual. Thus the sets preserve their angular relations to each other while striking in abnormal directions. It is unusual for all the sets to be present in a single outcrop.

At most localities close to the western escarpment many of the surfaces in one or more sets are open or partly filled fissures. This dilatation of previously

Volume of Allochem components		>10% allochems		<10% allochems			
		Sparry calcite cement > microcrystalline ooze matrix	Microcrystalline ooze matrix > sparry calcite cement	1–10% allochems		<1% allochems	Undisturbed reefal rocks
		Sparry allochemical rocks	Microcrystalline allochemical rocks				
>25% Intraclasts		Intrasparrudite Intrasparite	Intramicrudite* Intramicrite*	Most abundant allochem	Intraclast-bearing micrite*	Micrite; if disturbed, dismicrite	Biolithite
<25% Intraclasts	>25% ooliths	Oosparrudite Oosparite	Oomicrudite* Oomicrite*		Oolite-bearing micrite*		
<25% Intraclasts, <25% Ooliths, Volume of fossils: pellets	>3:1	Biosparrudite Biosparite	Biomicrudite Miomicrite		Fossiliferous micrite		
	3:1-1:3	Biopelsparite	Biopelmicrite		Pelletiferous micrite		
	<1:3	Pelsparite	Pelmicrite		Pelletiferous micrite		

Table 2 Simplified Folk classification of non-dolomitised limestones
* = rare rock type

closed fractures probably occurred during the late-Pleistocene superficial movements as a consequence of the flow of the underlying Liassic clays into the vales.

Throughout this text Murray is responsible for the stratigraphy and palaeontology and Hancock for the structural data.

An introduction to the Folk classification of limestones. The term limestone embraces all the calcareous sedimentary rocks. This short note is a précis of the classification proposed by Folk (1959, 1962).

Basically limestones are made up of three components: complex carbonate particles (allochems); microcrystalline calcite ooze; sparry calcite cement.

The important allochems are: fossils and shell debris (bioclasts); pellets (spheres or ellipsoids with a microcrystalline internal structure); ooliths (spheres with concentric and radial internal structure) and intraclasts (fragments of penecontemporaneous limestone).

The microcrystalline calcite ooze is termed micrite. It probably originated as aragonite mud but has now recrystallised to calcite.

Sparry calcite cement consists of grains or crystals of calcite, usually more than 10 microns in diameter, growing from the allochems into the pore spaces.

In the Folk classification the limestone name is built up from its components. Thus a rock rich in ooliths and having a sparry calcite cement is an *oosparite.* If the ooliths are large it is an *oosparrudite.* Any coarse grained limestone has *rud* added to its name. *Bio* refers to bioclasts, *intra* to intraclasts, *pel* to pellets, *micr* to micrite. The classification of non-dolomitised limestones is shown in Table 2.

Itinerary

The following itineraries are proposed with different objects in mind. In each case the number of localities to be visited will take two days or more depending on the degree of detail of examination.

To see the stratigraphic succession:
2 Wotton Hill; **9** Leigh's Quarry; **11** Quarry 4, Selsley Common; **14** Catbrain Quarry, Painswick.

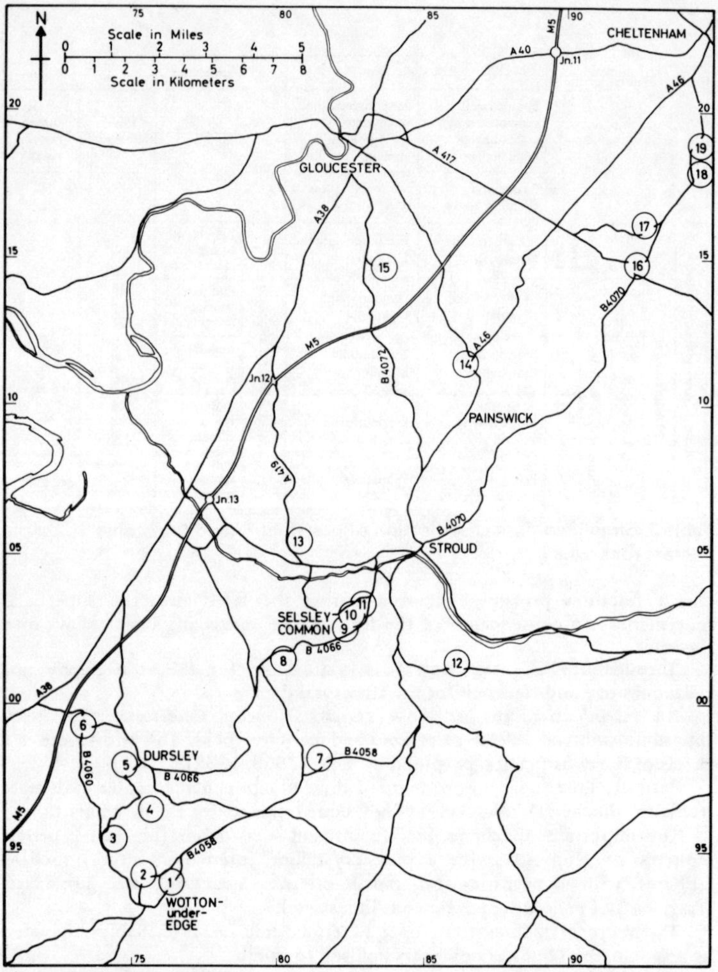

Fig. 1 Cotswold localities

Alternative route:

1 Waterworks and Coombe Hill Quarries, Wotton-under-Edge; 3 Nibley; 19 Leckhampton Hill; 18 Salterley Grange.

To study the Cephalopod Bed:

2 Wotton Hill; 3 Nibley.

To study the Lower Inferior Oolite:

1 Waterworks and Coombe Hill Quarries, Wotton-under-Edge; 2 Wotton Hill; 3 Nibley; 4 Breakheart Hill; 5 Dursley; 9 Leigh's Quarry, Selsley Common, Quarries 2 (10) and 4 (11); 14 Catbrain Quarry, Painswick; 17 Crickley Hill; 19 Leckhampton Hill.

To study the Middle Inferior Oolite:

14 Catbrain Quarry, Painswick; 19 Leckhampton Hill; 18 Salterley Grange.

To study the unconformities:

2 Wotton Hill; 3 Nibley; 4 Breakheart Hill; 8 Stanley Wood Quarry; 9 Leigh's Quarry; 14 Catbrain Quarry, Painswick; 16 Birdlip; 19 Leckhampton Hill.

To study the Upper Inferior Oolite:

1 Coombe Hill Quarry; 3 Nibley; 4 Breakheart Hill; 5 Dursley; 8 Stanley Wood Quarry; 9 Leigh's Quarry; 14 Catbrain Quarry, Painswick; 16 Birdlip.

Quarries in other strata:

6 Stinchcombe: M. Lias; 13 Stonehouse: L. Lias; 15 Robins Wood Hill: L. & M. Lias; 12 Simmonds' Quarry: Great Oolite; 7 Pike Quarry: Great Oolite.

1 Waterworks and Coombe Hill Quarries, Wotton-under-Edge (ST 768 943 & ST 768 944) Leave Wotton-under-Edge on the B 4058 (Tetbury road); on the sharp bend of the road on the second spur of Coombe Hill is the Waterworks Quarry of Ager & Donovan (in McKerrow *et al.* 1964). There is easy access from the road and cars or coaches can be parked in the quarry.

The succession extends from the Lower Inferior Oolite (Lower Limestone) to the Upper Inferior Oolite (Upper Trigonia Grit, Fig. 2). The Lower Limestone is a massive oosparite with abundant shell debris (echinoderm, molluscan and polyzoan) along certain depositional planes. Small scale cross-bedding is shown on some faces. The 30 cm of rubbly oosparite which follows has been tentatively equated with the Pea Grit by Ager & Donovan (1964). The ooliths and echinoderm bioclasts are sutured at their contact, suggesting solution. Following this are some 6.7 m of Lower Freestone (oosparite with echinoderm and polyzoan bioclasts) which is more conspicuously bedded than the Lower Limestone. Some of the bedding planes have reddish-brown marly partings. This part of the succession can be best studied by ascending the grassy face along the wide gull which cuts it.

The upper part of the Lower Freestone is much shattered, probably by frost action. At the top is the Middle Bajocian unconformity. Ager & Donovan (1964) note the presence of borings in the topmost Lower Freestone but they are not obvious at the time of writing. Instead there is a prominent calcite vein along the unconformity in the central part of the quarry. The succeeding 1.1 m of Upper Trigonia Grit are of Upper Inferior Oolite age. The rock is a biomicrite with abundant fossils, notably *Trigonia* and *Acanthothyris*.

Coombe Hill Quarry may be reached by walking along the road up the hill for about 100 m.

The rocks here are both topographically and stratigraphically higher than those of the Waterworks Quarry. They represent the Clypeus Grit. The section is obviously less complete than when Richardson described it (1910, pp. 103–4). The lowest bed which may be exposed forms a small ledge at the base of the SW–NE face. It is an oosparite which is strongly bored on its upper surface. This is probably the lowest horizon within the Clypeus Grit. The next bed exposed in the overgrown bank at the foot of the cliff is an intramicrite with *Stiphrothyris* (=*Terebratula globata* of earlier authors) and *Acanthothyris* – about 35 cm thick.

The main SW–NE face consists of biomicrite at the base followed by an alternation of thin soft rubbly partings and hard intramicrites. All are rich in *Stiphrothyris* and *Acanthothyris*. The intraclasts (representing eroded fragments of a pre-existing limestone) are more weathered than the matrix because this is their second involvement in a weathering/erosional cycle. Their importance is that they show that during Clypeus Grit time earlier-formed limestones were being eroded adjacent to where the Clypeus Grit was

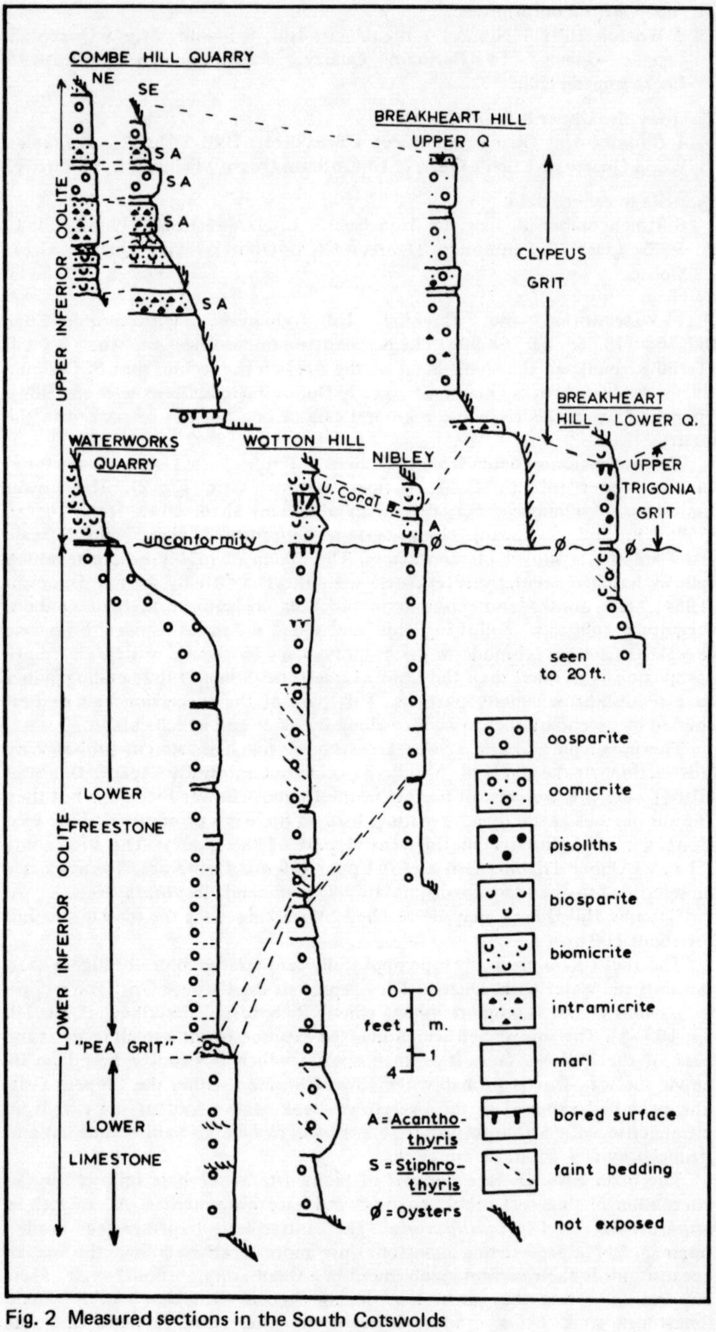

Fig. 2 Measured sections in the South Cotswolds

accumulating. Further, the association of the brachiopods *Stiphrothyris* and *Acanthothyris* with the intramicrite suggests that they originally lived attached to the rock which was being eroded. None of the brachiopods have been found to be preserved in the life position but as they are so well preserved it seems unlikely that they have been transported far after death. The uppermost horizon exposed at present is an oosparite.

The section on the SE–NW face differs somewhat from that of the face already described (Fig. 2). In particular the relationship between the intramicrites and the intervening rubbly partings is different and there is some variation in bed thickness.

Richardson mistook the intraclasts for ooliths and described the upper part of the Clypeus Grit as the White Oolite by analogy with the White Oolite at Horton to the south. The Wotton area is transitional between the characteristic Upper Bajocian successions of the north and south Cotswolds (Table 1) and the White Oolite was regarded as an intermediate facies.

2 **Wotton Hill** (ST 753 938) Branching off the B 4060 to the west of Wotton-under-Edge is the Old London Road. This climbs up Wotton Hill. About half-way up the hill, opposite a small lane which leads down to The Butts, is an old quarry on the west side of the road hidden in the trees. This shows the finest section of the Cephalopod Bed in the area; it is also one of the two type sections (the other being on Frocester Hill).

The quarry can be divided into three units. There is a central section with a long scree at its base; this is the one recorded in Fig. 2. To the left of the scree (SW) the whole succession has slipped down hill; some 6 m of Cotteswold Sands are exposed. To the right of the scree (NE) the Cotteswold Sands are again exposed and the succession can be traced without break from here up through the central face.

The Cotteswold Sands of Upper Lias age are micaceous fine-grained sands and silts. They are followed by a series of condensed deposits known collectively as the Cephalopod Bed due to the prevalence in them of both ammonites and belemnites as fossils. Richardson (1910, pp. 105–6, Wotton Hill Quarry II) described the succession in detail and related it to the ammonite zones. The section has been remeasured (Fig. 2).

The basal part of the Cephalopod Bed is a 30 cm thick band of biomicrite-biomicrudite (the Striatulum Bed). A little quartz sand is present. Echinoderm debris is abundant and some of it is recognisable as *Pentacrinites*. Chamosite infills the stereome of some of the echinoderm material. The abrupt termination of the supply of clastic material at the top of the Cotteswold Sands is most marked.

The Pedicum Bed is a 25 cm soft ferruginous marl with chamosite ooliths. It is followed by the Strickmani Bed, 30 cm of very sandy, ferruginous, biomicrite. This rock contains flattened chamosite ooliths and, locally, chamosite infills the stereome and also partially replaces the echinoderm debris. The next 1.88 m of soft ferruginous marls with chamosite ooliths was subdivided by Richardson into three beds spanning the dispansum and levesquei subzones. This is capped by 25 cm of harder ferruginous marl with chamosite ooliths. Above this the soft ferruginous marls with chamosite ooliths (48 cm) were subdivided by Richardson into three beds spanning the moorei and aalensis subzones.

Then follows a series of harder beds which form the upper part of the cliff in the middle section of the quarry. The upper part of this succession, the Scissum Beds, represents the base of the Inferior Oolite. The limestones of the Scissum Beds are sandy biomicrites and mark the renewal of the supply of clastic sedimentary material for a brief period.

Much of this succession yields ammonites, belemnites, bivalves (particularly burrowing forms such as *Pholadomya*) and brachiopods (notably

Homoeorhynchia cynocephala).

The higher quarry on Wotton Hill (= Wotton Hill Quarry 1 of Richardson 1910, p. 104) may be reached by walking through the woods or by returning to the road and following it up the hill. This quarry shows a section through most of the Lower Inferior Oolite freestones, the unconformity and part of the Upper Inferior Oolite (best seen in the southern part of the quarry).

The Lower Limestones of the Lower Inferior Oolite are readily accessible along the foot of the cliff throughout the quarry. They are oosparites with local lenses of bioclastic material (echinoderm, bivalve and polyzoan) along some of the depositional planes. Quartz sand is present to perhaps 10%. Cross-bedding is sometimes seen and in general the foresets point towards the NE. Although the Lower Limestones are separable into a number of beds, the bedding planes are merely breaks and are free of the marly material which characterises those of the Lower Freestone.

The Pea Grit is a rubbly limestone horizon 15 cm thick separating the two major freestone units. The Lower Freestone is an oosparite with abundant echinoderm debris and quartz sand may form 10% of the rock. The bedding planes characteristically have a thin marly parting. Cross-bedding is evident at several levels (Fig. 2). At 109 cm below the unconformity is a bored surface within the Lower Freestone. The unconformity is excellently displayed. The top of the Lower Freestone is bored to a depth of 25 cm by annelids and by the bivalve *Lithophaga*. Above follows 106 cm of Upper Trigonia grit separable into two beds; a lower one of intermediate biomicrite/biosparite with echinoderm and bivalve bioclasts and some quartz sand grains, and an upper more massive bed of similar composition but lacking quartz sand grains. The top of the upper bed is extensively bored to a depth of 30 cm.

The Dundry Freestone equivalent of Richardson is not really evident now. The highest beds are attributed to the Upper Coral Bed. These are lenticular masses of intermediate biomicrite/biosparite.

3 **Nibley Knoll** (ST 745 957) The Nibley section has long been known to geologists. The first detailed description was that of Buckman (1888, in his Monograph, 1887–1907, pp. 46–7; and 1889, p. 445) and this was followed by a more detailed description of the Nibley Knoll Quarry section by Richardson (1910, p. 107).

Cars and coaches may be parked in the lay-by opposite the lane leading to Tyndale Monument (ST 740 956). This lane is deeply incised in the Cotteswold Sands and gives the best section available in the southern Cotswolds at present. About one fifth of the way up the hill, opposite the steps leading to Tyndale Monument, an old working on the left of the lane shows two hard fine-grained siltstone bands each about 23 cm thick separated by 60 cm of softer silty sand. The hard bands pass laterally into about 1.8 m of hard siltstone. These harder rocks have yielded a good fauna of belemnites and ammonites (particularly *Haugia* spp.).

Proceeding up the hill further sections of soft silty, very fine grained sands (Cotteswold Sands) may be seen. At the top of the high bank on the right of the lane hard bands are seen to appear, and as the path rises these bands intersect it about 50 m from the entrance to the quarry. These beds are the Cephalopod Beds, the condensed sequence at the top of the Upper Lias. The exposures are poor compared with those on Wotton Hill and it is not readily possible to make out the succession in detail. However, for the collector there are usually good finds to be made in the screes beneath the outcrop; belemnites, ammonites and bivalves are usually common.

Sandy limestones of the Scissum Beds are exposed immediately outside the entrance to the quarry in the floor of the lane while in a small shallow, overgrown quarry to the left (ST 745 958) there are some 1.8 m of cross-bedded Lower Limestone.

The main quarry on Nibley Knoll (ST 745 957) shows a section through Lower Inferior Oolite, the Bajocian unconformity and about a metre of Upper Inferior Oolite (Fig. 2). The lowest beds exposed are the cross-bedded oosparites of the Lower Limestone seen in the lower part of the NW–SE face. There is great variation in grain size due to the presence of laminae of coarse ooliths and shell debris in an otherwise medium-grained rock. The foresets of the large scale cross-bedding show a general trend to the NE.

Following the Lower Limestone are 5 cm of rubbly limestone which have been regarded in the past as the equivalent of the Pea Grit. Above comes the Lower Freestone, an oosparite with laminae of coarse ooliths and shell debris.

At the top of the Lower Freestone is the unconformity surface represented by borings of annelids and *Lithophaga* extending down 20 cm from the oyster-encrusted upper surface. This unconformity marks the absence of rocks of Middle Inferior Oolite age.

Immediately above the unconformity is a marly parting followed by 40 cm of richly fossiliferous biomicrite yielding *Acanthothyris* and moulds of *Trigonia*. This is the Upper Trigonia Grit. The 15 cm of rubbly fossiliferous limestone above this represents the Upper Coral Bed. The uppermost part of the quarry face is biomicrite of the Clypeus Grit. Serpulids are abundant and intraclasts are present in the biomicrite matrix.

The quarry can be left via a gate in the eastern corner. It is worth walking to the foot of the Tyndale Monument in order to admire the view. In particular the shape of the escarpment is clearly shown. The Inferior Oolite forms the plateau at the top of the hills, the Cotteswold Sands form the steep face of the escarpment. The Middle Lias Marlstone forms the platform at the foot of the Hill along which the road runs. Wotton-under-Edge, North Nibley, and Stinchcombe are built on this platform. Beyond it is another escarpment leading down to the low land of the Lower Lias.

From here one can descend the hill via the lane (the steps have for some years been disrupted by landslides) or one can walk through Westridge Wood to the Wotton Hill quarries.

4 **Breakheart Hill Quarry** (ST 756 967) This quarry can be approached from North Nibley by taking the small road down to Millend and then ascending the steep hill (1 in 4) up Breakheart Hill, or it can be approached from Dursley. It is necessary to obtain permission to enter this quarry (see p. 140).

The section was originally described by Woodward (1894, p. 108) and noted by Richardson (1910, p. 109). However, the quarry was re-opened some years ago to provide stone for the nuclear power station at Berkeley. The newly enlarged section was described by Murray (1969).

In the central, deepest part of the quarry, about 5.5 m of Lower Freestone may be seen in the NE corner. It is divided up into beds generally less than 30 cm thick separated by softer marly and somewhat rubbly partings. Some beds are clearly cross-bedded. Much of the cliff is in a dangerous condition and should be approached with caution. In the SW part of the deeper quarry the rotational effects of cambering are clearly displayed.

Passing now to the platform above the deeper quarry it can be seen that this is the unconformity surface. All the beds above this have been quarried out or pushed to one side. The unconformity is affected by the many gulls associated with the cambering. In general the effect has been for the cambered blocks to rotate about an axis parallel with the hillside (NW–SE), with a downward movement downslope (SW) and an upward movement upslope (NE). The result is a series of tilted blocks showing small scale 'escarpments' a few centimetres high facing uphill (NE). In other quarries the unconformity is marked by the presence of annelid and *Lithophaga* borings into the underlying limestone or by the presence of oysters adhering to the upper surface of the limestone. In this quarry borings are not very obvious but the relationship of the oysters to the unconformity is clearly seen; they show a marked clumped distribution – high density patches showing a wide range of oyster size (and therefore stage of development) are separated by extensive areas of bare rock free of oysters. What we see must closely resemble the sea floor of early Upper Trigonia Grit time. The freestone immediately beneath the unconformity is a cross-bedded biosparrudite with echinoid and polyzoan debris predominating.

Above the east face of the deeper quarry a small unquarried remnant of the Upper *Trigonia* Grit rests on the unconformity (Fig. 2). The first ledge with scree on it above the deeper quarry face is within the freestone. Above the scree is seen the cross-bedded biosparrudite which underlies the unconformity. The latter is marked by oysters overlain by 7 cm of softer oolitic and bioclastic limestone. The next 1.2 m of hard oobiosparite represents the Upper Trigonia Grit. This rock contains intraclasts and pisoliths with *Girvanella* in addition to columnals of *Pentacrinites* and polyzoan fragments. Many of the bioclasts have micrite envelopes. The upper surface of this bed is bored; this probably represents a small local unconformity or non-sequence. The top 38 cm of broken-up limestone is highly fossiliferous (terebratulids and *Ctenostreon*) and probably represents the lower part of the Clypeus Grit. Elsewhere in the quarry this part of the succession has either been removed or is at present obscured by bulldozed debris.

The upward continuation of the Clypeus Grit succession can be seen in the NW part of the quarry. The lowest bed exposed forms a terrace from which the overburden has been removed. It is an intramicrite, 20 cm thick, with abundant *Stiphrothyris* and *Acanthothyris.* Two loose blocks believed to have come from this horizon have yielded *Parkinsonia parkinsoni.* The adjacent quarry section is about 4 m high (Fig. 2). Above a small scree, 1.5 m of oosparite (with intraclasts and *Stiphrothyris*) are followed by a soft parting, then two further beds of oomicrite with *Stiphrothyris* and *Acanthothyris.* These are separated from the topmost bed of oomicrite by 15 cm of rubbly limestone and the succession is capped by a rubbly biomicrite with *Stiphrothyris.*

Because so much of the Clypeus Grit has been stripped from the quarry, there are excellent horizontal, near-bedding plane, exposures and this quarry is consequently the best locality in the Cotswolds for collecting from this formation.

5 Dursley (ST 749 978) The numerous workings around Stinchcombe Common are now so overgrown that they are not worth visiting. However, on the north side of the Broadway, the road leading from Dursley WSW to Stinchcombe Common, a fine section through the Inferior Oolite remains. The section was described by Murray (1969) and it was mentioned briefly by Woodward (1894, p. 108).

The quarry has been worked at two main levels; an upper one with an entrance from a lane branching off to the NE close to the top of the hill, and a lower one which opens directly on to the road. Cars and coaches may be parked in the lower quarry.

The lower quarry is cut in Lower Limestone and Lower Freestone. The Lower Limestone, an oosparite, is best seen on the east face (Fig. 3) where two massive beds occur. The lower one shows small scale cross-bedding. The central part of the quarry is badly affected by gull development and cambering. The second section was measured to the west of the large gull with the cave developed in it. Here some 3.3 m of massive oosparite with echinoderm and polyzoan bioclasts are followed by 23 cm of marly, nodular oomicrite which may be the Pea Grit equivalent. The rock includes pisoliths, echinoid and polyzoan bioclasts and grapestones. Above this the Lower

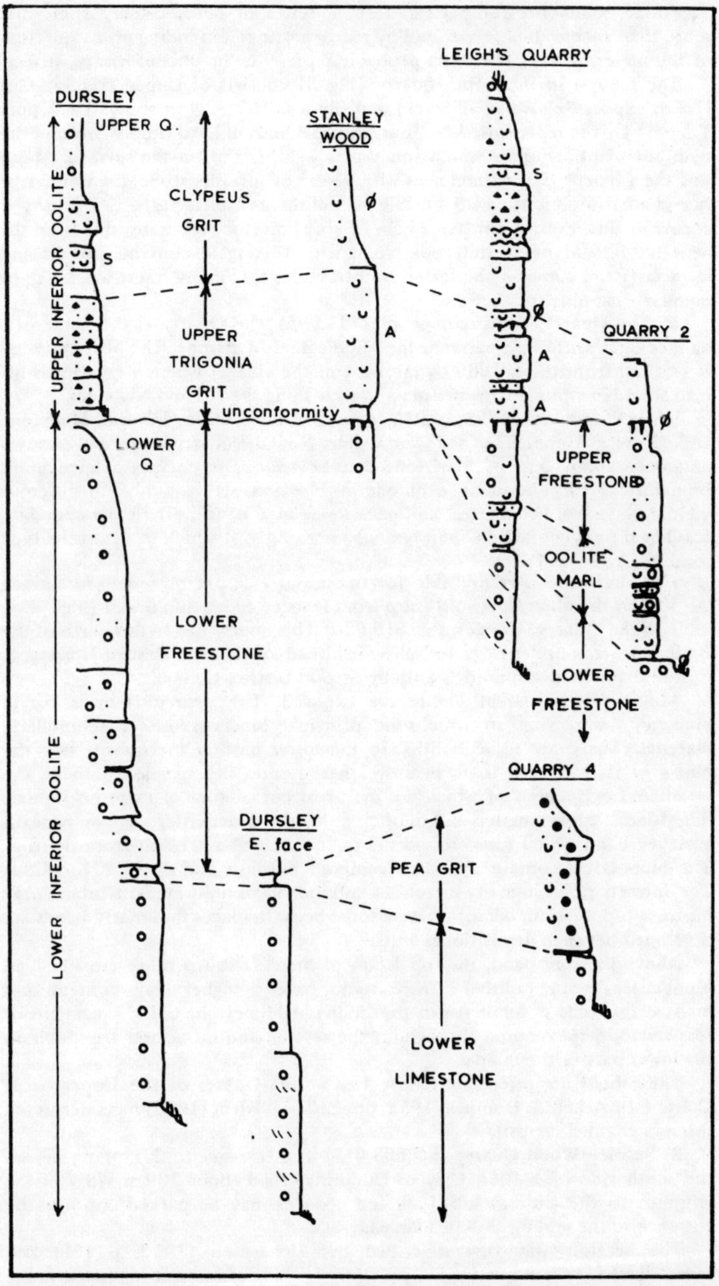

Fig. 3 Measured sections in the Mid Cotswolds (see Fig. 2 for key to ornament)

Freestone, oomicrites/oosparites, form a series of beds, usually somewhat more than 60 cm thick, separated by marly partings, extending up to the floor of the upper quarry. The latter probably represents the unconformity surface.

The section in the upper quarry (Fig. 3) consists of Upper Trigonia Grit (1.7 m exposed above small scree) and Clypeus Grit (2.4 m seen). The Upper Trigonia Grit is represented by four principal beds of hard rubbly intramicrite with abundant serpulids which sometimes weather out on the surface. Above this the Clypeus Grit commences with 38 cm of soft limestone, then 30 cm of fine-grained biomicrite with occasional ooliths and intraclasts. *Stiphrothyris* occurs at this level. A further 25 cm of soft limestone separates this from the next hard band of fossiliferous biomicrite. This again contains ooliths and intraclasts and some of the latter are of oomicrite. Finally, there are 1.3 m of oomicrite capping the section.

6 The Quarry, Stinchcombe (ST 735 995) The Quarry is the name of a hamlet built within a quarry in the Middle Lias Marlstone. The best exposure is reached from the B 4066 to the west of the village, where a gate leads off into a field in which exposures may be seen along the NE and NW sides.

When Woodward (1893, p. 215) described the section, 4.6 m of Marlstone and 45 cm of Upper Lias Clay were visible. Now the quarry is more overgrown and only about 3 m of Marlstone can be seen. The rock is a ferruginous biomicrite with abundant echinoderm bioclasts. It is richly fossiliferous yielding bivalves, belemnites, and occasional large nautili. The most abundant fossil is the rhynconellid *Gibbirhynchia micra* Ager of which this is in the type locality (Ager 1954).

The Marlstone is responsible for the minor escarpment and terrace along the foot of the main Cotswold scarp from Dursley to Wotton-under-Edge.

7 Pyke Quarry, Horsley (ST 815 980) This quarry lies to the north of the B 4058 Wotton-under-Edge to Nailsworth road about 3 km west of Horsley. It is now abandoned and used as a tip by Stroud District Council.

About 6 m of Great Oolite are exposed. The quarry floor is cut in blue-grey biomicrites in which the principal bioclasts are of echinoderm material. There are some ooliths. In the lower part of the quarry face the blue-grey rock is seen to form unweathered cores in larger joint-blocks, the weathered peripheries of which are the usual buff colour of exposed Jurassic limestones. Approximately 4.3 m of cross-bedded biomicrites are seen beneath a clayey band which runs around the quarry face about 1.8 m from the top. The biomicrites contain abundant echinoid bioclasts and up to 20% ooliths. The foresets point in many directions and individual units are lenticular. Large burrows (up to 1 cm diam.) traverse some beds. In places thin marly bands are developed between depositional units.

Above the clay band, the top 1.8 m of the cliff shows more cross-bedded biomicrites (with ooliths). These rocks have a higher clay content and marly-clay bands occur between thin individual limestone units. Fragments of fossil wood are common throughout the section and intraclasts are visible in the lower part of the quarry.

These beds are referrable to the Hen's Cliff Oolites of the Upper Great Oolite (see Arkell & Donovan 1952, pp. 240–1). Klein (1965) has interpreted this as a channel deposit.

8 Stanley Wood Quarry (SO 803 017) The entrance to this quarry lies on the south side of B 4066 Uley to Dudbridge road about 300 m WSW of the entrance to the gliding club. Cars and coaches may be parked opposite the entrance to the gliding club in a large lay-by.

The section was first described by Richardson (1910, p. 121) and redescribed by Channon (1950, pp. 256–7). It is of interest in that it is the only accessible exposure in the rocks above the unconformity between Dursley (the Broadway Quarry) and Selsley Common (Leigh's Quarry). The succession (Fig. 3) commences with 0.9 m of oosparite with little bioclastic material and free macrofossils. Richardson (1910, p. 121) regarded this as Lower Freestone but Channon (1950, p. 256) believed that it is Upper Freestone on evidence obtained from the nearby Marmontsflat Wood Quarry (SO 803 018). However, on lithological grounds it closely resembles the Lower Freestone of Leigh's Quarry on Selsley Common.

The unconformity is marked by a bored surface in the top of the freestone. Above this are 2.34 m of Upper Trigonia Grit, the lower 20 cm being softer than the rest. The rock is a massive fossiliferous biomicrite but also includes biosparite and contains *Acanthothyris, Trigonia* and *Trichites.* The brachiopods are sometimes infilled with calcareous *Rhaxella* spicules (see Sorby 1851 and Wilson 1966).

The Clypeus Grit is a fine-grained biomicrite with oysters and terebratulids. This rock has been shattered by frost action. Channon (1950) regarded this bed as Doulting stone rather than Clypeus Grit.

9 **Leigh's Quarry, Selsley Common** (SO 826 025) This quarry is on the edge of Selsley Common north of Selsley Hill Farm.

This is one of the classic sections in the Inferior Oolite. It was first described by Witchell (1882) but a more detailed record has been provided by Richardson (1910, pp. 125–6). A more complete section of Inferior Oolite is shown here than at any other quarry in the South Cotswolds (Fig. 3).

The lowest beds exposed are the massive oosparites, showing large scale cross-bedding, of the Lower Freestone. The Oolite Marl commences with 5 cm of soft marl. Then follow two beds of fine-grained biomicrite, the upper containing scattered ooliths. In the parting between these beds, Richardson noted small sponges. The limestones yield a fauna of brachiopods.

The Upper Freestone commences with a thin marly parting followed by 1.2 m of oobiosparite. The top is bored by annelids and *Lithophaga* and this represents the unconformity. Thus, the unconformity here occurs at a much higher level within the Inferior Oolite than it does further south (see Table 1).

The Upper Trigonia Grit is represented by 1.7 m of biomicrites all of which are fossiliferous. The basal bed was separated off by Witchell as the Gryphite Grit.

The Clypeus Grit commences with a marly parting with oysters resting on top of the Upper Trigonia Grit and this probably represents a non-sequence. Following this a 79 cm bed of somewhat irregularly bedded biosparite passes up into a 23 cm bed of mixed biosparite and biomicrite. The upper surface of this bed and the upper and lower surfaces of the succeeding bed are very uneven and the rock becomes a biomicrite with intraclasts. Above this are 69 cm of intramicrite with fossils including *Stiphrothyris.* In the upper two beds the intraclasts are less common but are accompanied by ooliths in an essentially biomicrite rock. These beds are fossiliferous and again yield *Stiphrothyris.*

From the grassy bank above the cliff face Richardson recorded debris of the White Oolite and pieces may still be found. The rock is an oosparite with intraclasts.

In the nearly horizontal limestones at this quarry, surfaces belonging to vertical minor fracture sets J_1 to J_4 are orientated in the usual manner. The average strikes of sets J_1, J_2, J_3 and J_4 are 051°, 129°, 178° and 087° respectively. Wrench shear sets J_1 and J_2 enclose an average angle of 78° about an E–W bisectrix. Extension joints and veins in sets J_3 and J_4 bisect the angles between the wrench shear surfaces. A few of the J_2 shears bear a penetrative lineation plunging NW at about 5°. The majority of the fractures in the quarry belong to sets J_2 and J_3, and they are best developed in the Lower and Upper Freestone and the *Clypeus* Grit. Some members of all sets, but especially those belonging to sets J_2 and J_4, are fissured as a result of dilatation during the late-Pleistocene.

10 Quarry 2, Selsley Common (SO 829 032) If one proceeds across the common from Leigh's Quarry to the tumulus, the first quarry encountered (just beyond the tumulus) is Witchell's Quarry No. 1, which is now completely overgrown. Beyond this is Quarry No. 2 which still shows a reasonable section along the south face (Fig. 3).

Only a few centimetres of Lower Freestone are exposed at the top of the scree; it is an oosparite and on the upper surface are oysters. This may indicate a slight non-sequence before the Oolite Marl. The latter consists of four limestone bands separated by five marl bands, the lower one being about 10 cm thick. When washed on a 200 mesh sieve, the latter yields small rhynconellids, spines of echinoids and columnals of *Pentacrinites.* The limestone bands vary in thickness laterally. They are variable also in lithology; biomicrite, biosparite and oosparites with ooliths and bioclasts being unevenly distributed throughout the thickness of a single bed. Composite grains and intraclasts of oolitic rock are present in some of the beds. This part of the sequence is suggestive of fairly low energy conditions with ooliths being introduced periodically, perhaps by the action of storms. It could represent an outer lagoon environment.

The Upper Freestone shows a greater thickness here than at Leigh's Quarry. It is an oomicrite although in places the matrix becomes sparry. The ooliths are densely packed in this rock. Some have a core of a fragment of an earlier oolith – perhaps an intraclast. Others form composite ooliths. The top of the Upper Freestone is bored by annelids and *Lithophaga* to a depth of 25 cm. Oysters rest on this unconformity surface.

The overlying Upper Trigonia Grit is a highly fossiliferous biosparite, much shattered by frost action.

11 Quarry 4, Selsley Common (SO 832 035) This is the only quarry on the lower flanks of Selsley Common which still shows a reasonable exposure in the lower part of the Inferior Oolite.

Above the scree 1 m of Lower Limestone (an oosparite with some intraclasts) is seen although a greater thickness is exposed in an isolated pinnacle. At the top of this bed there are occasional borings. (Fig. 3).

The Pea Grit reaches an exposed thickness of 2.9 m. Two units are visible, each commencing with a hard base and passing up into a softer, rubbly upper part. True pisoliths are present and these, of course, are the structures which prompted the name 'Pea Grit'. The pisoliths at this locality are flattened and elongate and most have large bioclast nuclei of polyzoa, brachiopods or bivalves. The matrix is a biomicrite. This formation is renowned for its abundant fauna of brachiopods (terebratulids and rhynconellids), echinoids and gastropods.

12 Simmonds' Quarry, Burleigh (SO 862 014) This quarry, which can be entered via a gate from Minchinhampton Common, is owned by Mr. Simmonds of Minchinhampton who now uses it as a builder's yard. Permission to enter must be obtained from Mr. Simmonds (see p.140). The quarry is no longer being worked although it was when Channon first described it in 1950. Much of the cliff is in a very dangerous condition at the time of writing and the beds are much disturbed by gulling.

Succession based on Channon (1950, p. 251) with some modifications:

		m
4	Soil and rubble	1.07
3	'Planking' equivalent. A series of massive beds of oolitic limestone of varying texture, white, grey and yellowish brown, with fossils distributed throughout, the most fossiliferous portion (gastropods and bivalves) being a 15–30 cm bed of very white, softish, shelly limestone nearly at the top of the series.	3.2

2 'The Scroff'. Yellowish stony marl full of large *Liostrea hebridica, Lima cardiiformis,* with a few other fossils. 0.02
1 'Weatherstone'. A series of hard, massive, current-bedded, coarsely oolitic limestones, shelly, brownish or grey, with very thin sandy partings. Excellent building stone; seen 9.1

A large fauna, primarily of gastropods and bivalves, was listed from this quarry by Channon.

Only the Weatherstones can be readily examined. These are oosparites with varying amounts of bioclasts. On the west face, to the north of a prominent gull, a bioclastic limestone is clearly seen infilling a channel in the oosparite. The latter commonly contains ramifying cavities which have been interpreted by Donovan (in McKerrow *et al.* 1964) as representing sponges.

Arkell & Donovan (1952) concluded from the limited ammonite evidence that the Scroff and Planking are equivalent to the Cross Hands Rock of the South Cotswolds and the Fuller's Earth Rock of Somerset. Another exposure of these rocks is the quarry alongside the lane from Minchinhampton Common to Burleigh Court (SO 867 015). The entrance is at the cross roads at SO 866 016. The quarry is overgrown but is of interest in that it was the source of stone for Gloucester Cathedral.

13 **Stonehouse Quarry, Nr. Stroud** (SO 810 053) This disused quarry may be reached from the A 419 which passes through Stonehouse. It lies to the north of the railway line.

According to Ackerman & Cave (1968) an unbroken sequence of 48 m of Lower Lias Clay, Dyrham Silts and Marlstone Rock Bed is exposed. They point out that the beds dip 5° to the NE and that the lower beds cannot be traced beyond the quarry. It is believed that the whole mass of rock has moved as a landslide to which they have given the name the Brick Pit Slide. Some of the harder bands within the clays yield bivalves and fallen blocks of Marlstone are often richly fossiliferous.

14 **Catbrain Quarry, Painswick** (SO 866 115) To the north of Painswick on the A 46(T) take the turning to the southwest at Castle Godwyn. About 300 m up the lane at the sharp bend, a small track leads south to Catbrain Quarry. At present the quarry is rented from the Lord of the Manor by Mr. Gerrish. For permission to enter the quarry see page 140.

The section has previously been described in outline by Hancock (1966) and by Murray (1969). Both the Lower and Middle Inferior Oolite are present. That at the south end of the quarry is best shown and is described here (Fig. 4).

In the lower part of the cliff 6 m of oomicrites and oosparites, with some composite grains and densely packed ooliths, of the Lower Freestone are followed by a 20 cm marl band and a further 2.4 m of oosparites and oomicrites (with composite grains and pisoliths at the base of the bed). Above, approximately 1.2 m of softer poorly exposed beds may represent the lower part of the Oolite Marl. Following this are alternations of soft marls and somewhat harder oomicrites (with abundant bioclasts) having a total thickness of 2.1 m. The topmost bed is bored on its upper surface and represents the unconformity. On lithological grounds there is no reason to believe that any Upper Freestone is present here.

The Middle Inferior Oolite commences with the Lower Trigonia Grit. This is believed to comprise a basal bed of bioturbated marl, in which internal moulds of *Pholadomya* are not uncommon, and two distinctive beds of fossiliferous biomicrite. These have been weathered along joint surfaces in such a manner that the recrystallised bivalve shells project from the biomicrite matrix. Intraclasts are present in the upper bed. The Buckmani Grit is distinctive in having a basal fine sand bed, 60 cm thick. The upper surface is moulded to the very irregular base of the succeeding sandy biomicrite which in

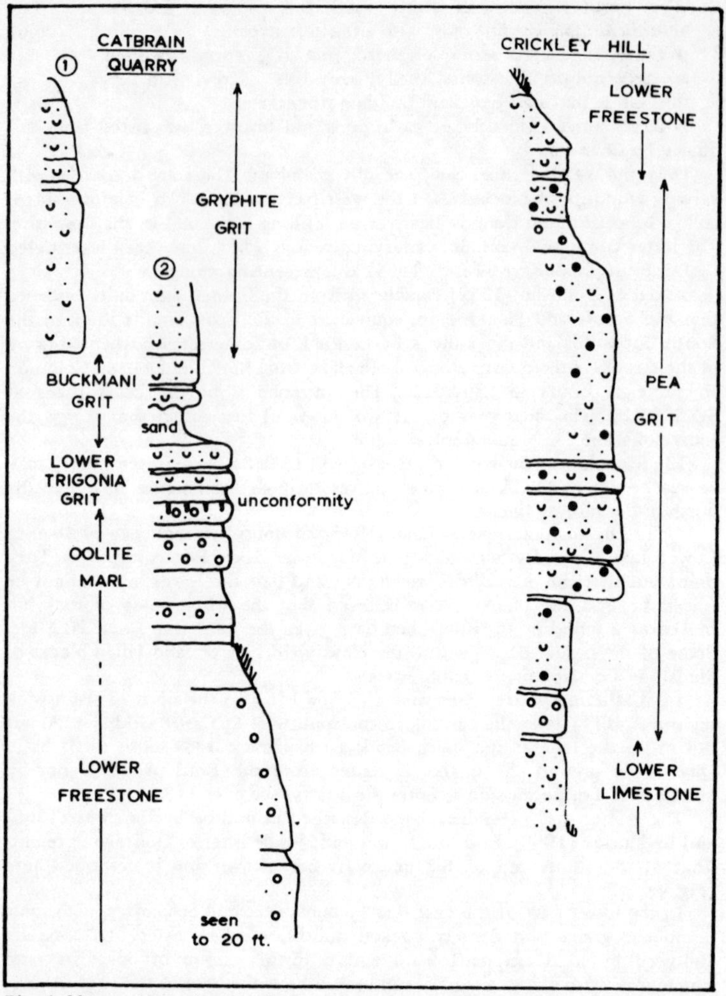

Fig. 4 Measured sections at Painswick and Crickley Hill (see Fig. 2 for key to ornament)

turn is separated by a thin marl band from a massive bioturbated sandy biomicrite crowded with bivalves. The remainder of the succession is Gryphite Grit. This is dominantly a rubbly, nodular, sandy biomicrite with irregular sandy partings. It is richly fossiliferous containing abundant *Gryphaea sublobata,* belemnites, and *Ctenostreon* at the base; it is extremely bioturbated.

There is some uncertainty about the boundaries between the Middle Inferior Oolite members here. Buckman (1895) described the Lower Trigonia Grit as 1.17 m of 'Yellow, ironshot, oolitic stone, somewhat marly' at Scotesquar Hill nearby, while at Kimsbury Castle it was stated to comprise 18 cm 'brown ironshot marl', 10 cm 'hard, brown ironshot shelly stone', and 2 cm 'brown marl'. At these two localities the sandy horizon within the

Buckmani Grit was 20 cm and 8 cm thick respectively and came above the base of the formation.

The importance of this exposure is that it lies in the Painswick syncline which preserves the Middle Inferior Oolite. These rocks are missing to the north at Birdlip and to the south at Selsley Common.

In the south and west faces of Catbrain Quarry it is possible to examine not only all six sets of minor tectonic fractures but also a variety of structures of superficial origin.

The fracture system, which is best developed and displayed in the Lower Freestone, has been tilted with the limestones which dip east at up to 30°. Using the technique of stereographic projection it is possible to investigate the geometry of the fracture pattern after the beds are imagined to be restored to the horizontal. After reorientation the average strikes of sets J_1, J_2, J_3 and J_4 are 033°, 124°, 172° and 081° respectively. These values indicate that the system is orientated about 10° anticlockwise from the usual arrangement. A few J_2 wrench shear surfaces bear slickenside striations on the wall rocks and vein calcite. After reorientation this lineation is approximately horizontal. Large joints belonging to normal shear sets J_5 and J_6 are particularly well developed in this quarry. After reorientation they strike approximately E–W and dip either north or south at about 55°.

Superficial structures of late-Pleistocene age include cambers, gulls, dip and fault structure, and, in the Oolite Marl, conjugate minor normal faults symmetrically orientated about the dip of the beds. The tilt imposed by cambering ranges from about 30° along the western face of the quarry to about 15° at the eastern end of the southern face. In the southern face two normal faults, inclined steeply to the west, displace the limestones for up to about 3 m, and reduce the effect of the easterly tilt. The fault in the centre of the face is also a major wedge-shaped gull which tapers downwards. Many fractures in sets J_1 to J_6 are dilated to form fissures or incipient gulls.

Throughout the Cotswolds many fractures are encrusted by thin superficial veneers of calcareous tufa or travertine. On some surfaces these encrustations are locally thickened into two sets of ridges, or calcite bands as they were called by Hollingworth *et al.* (1944). The bands, which are generally at right angles to each other, and about 1–2 cm in width, are sites where calcite precipitation was enhanced on the encrustations. One set of bands formed at approximately horizontal successive ground water levels, while the other set marks the vertical paths down which water preferentially seeped. The horizontal bands are generally longer, straighter and more pronounced than the vertical bands, which commonly merge downwards into the horizontal bands. Where they occur on surfaces from more than one set, the bands which were precipitated parallel to former rest-water levels, may be used to determine the present orientation of an initially horizontal band-defined surface. Thus if there has been tilting since the bands were formed, the band-defined surface will also be tilted.

In Catbrain Quarry bands of both types are present on fractures of all six sets. When the lineations, formed by the previously horizontal bands, are plotted on a stereographic projection they all lie on the cyclographic trace of a common plane. The attitude of this band-defined surface is within a few degrees of the present tilt of the beds. This indicates that there was little tectonic tilting before the bands were deposited, and that they ceased to form before cambering occurred. If they had continued to be precipitated during and after cambering it would be possible to determine the attitude of several band-defined surfaces.

On the west side of Painswick Beacon it is possible to see in a small quarry (SO 867 121) that the Freestones dip at about 30° to the WSW as a result of cambering. From a point on the upper rim of this quarry the westward cambering in it, and the eastward cambering in the Catbrain Quarry is visible.

15 Robins Wood Hill: Tuffley Old Brickworks (SO 836 149) Robins Wood Hill lies to the south of Gloucester. Take the B 4072 south from the Ring Road and after 0.6 km turn off to the SE along Fox Elms Road at Tuffley. This leads to the pits. Entry may be gained via a stile from this lane.

The succession here was described in detail by Ager (1956). Altogether he listed 17 beds but at present the lower four beds appear not to be exposed. Bed 5 is grey shales; bed 6, hard sandy shales; bed 7, sandy grey shales; bed 8, a highly fossiliferous ferruginous band from which *Androgynoceras* has been recorded. These beds constitute the *davoei* and *ibex* zones of the Lower Lias.

Following this are approximately 16 m of grey shales with bands of ferruginous nodules (bed 9), then a 30 cm ferruginous band (bed 10) and a further 60 cm of grey shales (bed 11). This part of the succession extends up to the floor of the upper quarry and is of uncertain age.

The upper quarry may be reached by a rough climb to the left of the main face or by a footpath round the south side of the pit. Following the grey shales of bed 12 is a prominent ferruginous, muddy, sandy biomicrite (bed 13) which yields a rich fauna of bivalves, *Amaltheus* and belemnites. More grey shales of bed 14 are followed by another prominent band of ferruginous siltstone (bed 15). These beds comprise the *margaritatus* zone of the Middle Lias.

Finally a thick sequence of grey shales (bed 16) are capped by the massive micaceous fine-grained siltstone (bed 17). This yields occasional specimens of the brachiopod *Gibbirhynchia micra* in the lower 15 cm. These two beds form the *Spinatum* zone of the Middle Lias.

16 Birdlip Hill (SO 924 144) On entering Birdlip from the south on the B 4070, turn left down Birdlip Hill. On both sides of the road there are sections and a little way down the hill at SO 925 146 is a much overgrown quarry.

The quarry section shows oomicrites, in which the ooliths are densely packed with sutured contacts, of the Upper Freestone. Above this is the unconformity surface which is bored and oyster-encrusted according to Hancock (1966).

In the road cutting on the north side of the road, beneath the grounds of the Royal George Hotel, the Upper Trigonia Grit and Clypeus Grit may be examined. Only 76 cm of Upper Trigonia Grit are exposed above the scree; the rock is a fossiliferous ferruginous biomicrite. A thin, soft parting separates this from the Clypeus Grit. There are two prominent massive beds, which are both 1.3 m thick, of intraclast-bearing biomicrite. Then follows a 28 cm bed of oomicrite in which there are pisoliths, grapestones and bioclasts, 30 cm of soft impure oomicrite and finally a hard biomicrite with pisolith-grapestones; in this rock many of the grains have a thin oolitic coating. The remainder of the succession is not accessible due to the steepness of the slope, but formerly the upper beds of the Clypeus Grit which underlay the grounds of the Royal George Hotel yielded many examples of *Clypeus ploti* from which the formation takes its name (Richardson 1904, p. 131).

17 Crickley Hill (SO 929 159) Access is from the A 417 Gloucester-Birdlip road. Cars may be parked at the entrance to the road leading to Cold Slad. Old workings occur on several sides of the hill but the one described here is that seen immediately above the junction with the road to Cold Slad, (Fig. 4) enter opposite Crickley Hill café car park.

Sparsely fossiliferous clays and sands attributed to the Upper Lias are exposed beneath the Inferior Oolite. The lowest Upper Lias is a greenish silky clay rich in mica flakes. Following this is ferruginous silty fine sand with doggers at its base. This contains a sparse ostracod fauna. Shelly lenses between the fine sand and the overlying blue shaly clay yield occasional ostracods and foraminiferids. No fossils which would be of value stratigraphically have been found in these deposits.

Above the blue clay is a band of ferruginous sandy biomicrite which represents the lowermost part of the Scissum Beds. In places the original blue-coloured unweathered cores can be seen in this biomicrite. Nodosariacean foraminiferids are quite common.

The main exposures in the Inferior Oolite in the old quarries are separated from the road cutting by a grass bank in which there are discontinuous exposures of the impure sandy biomicrites of the Scissum Beds.

The lowest bed in the quarry is a massive biomicrudite of the Lower Limestone. It is separated by a thin soft marl band from the next massive bed of oosparite with pisoliths which marks the base of the Pea Grit Series here.

Following this are four beds of pisolith-bearing biomicrite separated by thin marls and then a 3 m thick bed which is burrowed in the lower part and becomes coarsely pisolitic towards the top. The next 45 cm is an oosparite with scattered pisoliths, followed by 60 cm of rubbly pisolith-rich biomicrite which marks the top of the Pea Grit Series.

The uppermost exposed horizon is a hard biomicrite in which there are sheet-like masses of serpulids and many fragmented polyzoa. This bed may represent the base of the Lower Freestone.

18 Salterley Grange (SO 946 176 & SO 951 178) A small side road branches SE from the main B 4070 Cheltenham-Birdlip road at SO 944 178 and ascends the hill. On the left of this road is the large Salterley Grange Quarry. Cars may be parked in it. This quarry is in the Lower Freestone and the contact with the underlying Pea Grit (1 m exposed) may be seen on the east side.

Vertical wrench joints, veins and minor shear surfaces belonging to sets J_1 and J_2, striking at 052° and 152° respectively, are the most abundant fractures cutting the nearly horizontal limestones. Fractures in sets J_3 and J_4 are relatively rare in this quarry. Many of the J_1 fractures bear lineations plunging gently SW at up to 16° on the rock or vein surface. Although the lineations are concentrated mainly on fracture surfaces, some display slickolite characters and penetrate the rock up to 1–2 cm from the surface. Where there are lineations on J_1 veins they are commonly accompanied by small asymmetric accretion steps, orientated approximately perpendicular to the lineation. They indicate that there was oblique shear along the surfaces. Bedding planes are not noticeably offset across the shear surfaces despite this evidence for oblique slip.

Many of the fractures in set J_2 are fissured. The generally more closed aspect of J_1 surfaces, together with the fissuring of the J_2 surfaces, suggests that the local direction of underlying clay flow during the superficial movements was to the SW.

The WSW face of the quarry exhibits what appears to be a moderately inclined, north dipping, reversed fault striking approximately at right angles to the face. This structure is probably a vertical fault striking at a small angle to the face and downthrowing the rocks on its east side for about 1 m. The illusion of reversed slip is a consequence of the rocks on the downthrown side standing about 1 m east of the main face and being separated from it by a gently north sloping track.

At the top of the hill just beyond Hartley Cottages is a road junction. Turn left and immediately on the right is a large shallow quarry much of which is overgrown. In the NW corner a reasonably good exposure shows the following:

Upper Inferior Oolite

Upper Trigonia Grit – rubbly, bioturbated, highly fossiliferous biomicrite. 1.5 m

Middle Inferior Oolite

Notgrove Freestone – an oomicrite with abundant bivalve, polyzoan and serpulid bioclasts. The upper surface is intensely bored and represents an unconformity. 60 cm

Gryphite Grit – abundant *Gryphaea sublobata*, to be seen in the quarry floor.

This area is the type locality of the Upper Trigonia Grit (Donovan & Hemingway 1963, pp. 353–4).

19 Leckhampton Hill (SO 949 185) From the B 4070 Cheltenham-Painswick road take the turning to the SE along Daisybank Road. There is a car park on the right about 150 m from the main road. Take the footpath leading out of the SW corner of the car park and ascend the steep incline to the main quarry. (This is a public space owned by Cheltenham Council.)

Leckhampton Hill is one of the classic localities in the Cotswolds. It has figured prominently in descriptions of Cotswold geology from that of Murchison (1834) to that of Richardson (1904). Apart from the review of the Inferior Oolite by Arkell (1933), redescriptions by Murray (1969) and Macfadyen (1970), numerous reports of field trips, and a popular account by Dreghorn (1967), little has been added to our knowledge of the area from the time of Richardson (1904).

In the main quarry the section extends from the Upper Lias through the Lower Inferior Oolite with the unconformity in the Upper Freestone, to the Ragstones. However, the only part of the section, which is best examined in the main quarry, is the lower part up to the junction of the Pea Grit with the overlying Lower Freestone. The higher parts of the succession are not safely examined on the upper slopes of the quarry so alternative and better exposures are described in adjacent quarries.

The first section to be considered is that located between the old building and the concrete lime-kilns in the main quarry (Fig. 5). The Upper Lias Clay is kept exposed nowadays through the digging activities of visiting geologists. Then follows the sandy limestones of the Scissum Beds. These start with a 30 cm band of soft fine-grained sandstone which passes up into 1.67 m of impure muddy and sandy biomicrite which contains numerous pockets of soft ferruginous fine sand. Brachiopods are the dominant macrofauna but many of the bioclasts are of echinoderms and bivalves.

The base of the Lower Limestone is a hard bed of ferruginous biomicrite (abundant echinoderm bioclasts recognisable as crinoids) with ?chamosite ooliths. The succeeding 1.5 m of softer and more rubbly limestones are impure biomicrites again with abundant crinoid bioclasts. The Pea Grit is a series of limestones rich in fossils (mainly brachiopods and echinoids) and contains the pea-like pisoliths. The first bed with pisoliths is a biomicrite with bivalve and crinoid bioclasts. Some of these form the nuclei of the elongate and often flattened pisoliths in which the tubes of *Girvanella* can be seen. A thin (2 cm) marl band separates this bed from a series of hard beds which form a 1.8 m cliff. These are biosparrudites with pisoliths. Crinoid bioclasts predominate. Above is a grassy slope representing about 3.6 m of unexposed succession although in the cliff behind the concrete lime-kilns it is seen to be ferruginous rubbly limestone. The top of the Pea Grit consists of about 75 cm of oosparite enclosing large pisoliths. Many of the ooliths have a core of an echinoderm bioclast and complete columnals of *Pentracrinites* are present in the rock. The pisoliths show a complex history of development and enclose both bioclasts and ooliths within the spongy mass of *Girvanella* tubes. The pisoliths are very compressed oblate spheroids with the greater diameter generally parallel to the bedding/depositional surface. Many show preferential growth on one side

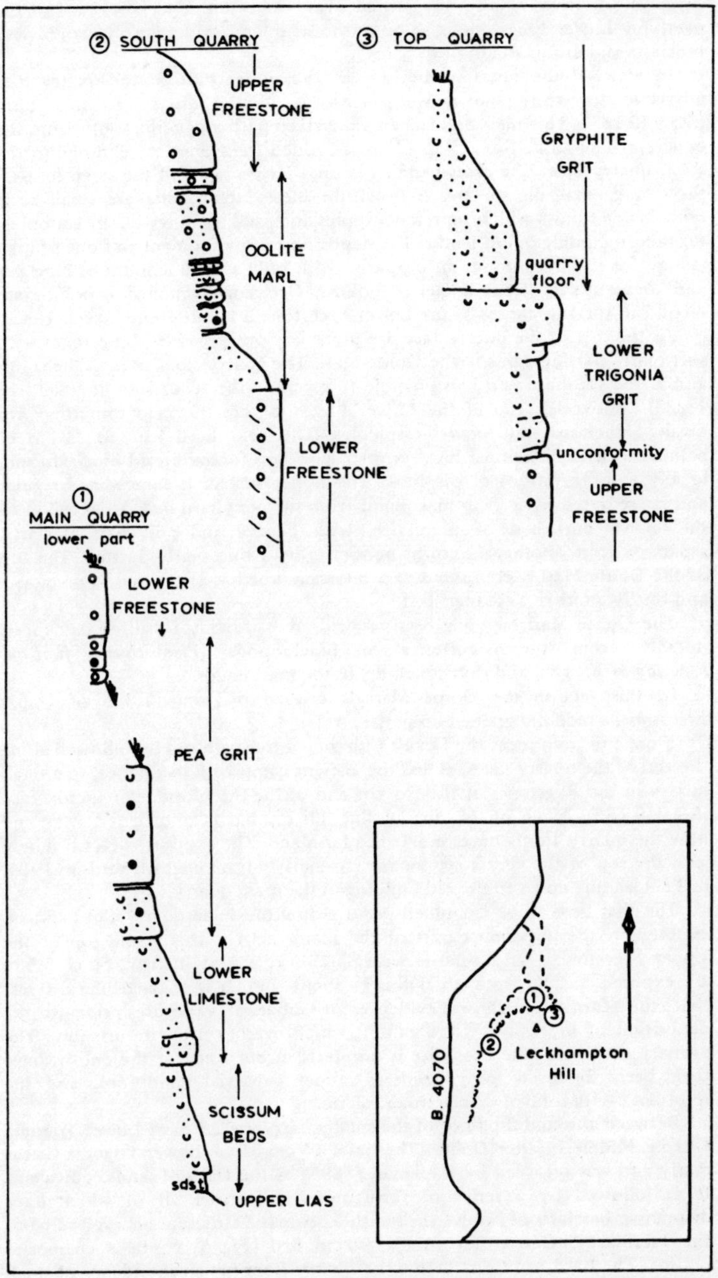

Fig. 5 Leckhampton Hill. Sketch map and the 3 measured sections (see Fig. 2 for key to ornament)

suggesting that they were not turned over too often. The basal part of the overlying Lower Freestone is an oosparite with large crinoid, polyzoan bivalve bioclasts and intraclasts of oomicrite.

In view of the unsafe condition of the upper part of the quarry it is advisable to study the Lower Freestone in the quarry to the south (SO 946 183). This may be reached via a steep path ascending south from the concrete lime-kilns. Some 9 m of cross-bedded freestones are exposed in the sheer quarry face. The cross-bedding is on a large scale and the steep foresets show a general dip to the SW. Within these larger units are small scale cross-bedded units which represent ripples and small sand waves (megaripples) on the prograding oolith banks. The depositional environment was one of high energy for the freestone is an oosparite with only a small amount of bioclasts and these usually form nuclei to ooliths. Occasional composite ooliths are seen. The total thickness of the Lower Freestone is about 40 m (Arkell 1933).

At the top of the quarry face are more obviously bedded limestones with soft marly partings; this is the Oolite Marl. The quarry floor rises to the south and at the southern end it is possible to examine the succession in detail (see Fig. 5). Above the top of the Lower Freestone are 90 cm of micrites with bivalve bioclasts and *Rhaxella* spicules. The next hard band is 40 cm of pelmicrite with occasional bivalve, gastropod and foraminiferid bioclasts and, locally, concentrations of pisoliths. The upper surface is somewhat irregular and is separated by a 2 cm marl band from the next hard bed. Then follows a thicker (60 cm) bed of oomicrite, with bivalve and polyzoan bioclasts, separated from another 43 cm of oomicrite by a thin marly parting. The top of the Oolite Marl is composed of a 5 cm marl bed, a 23 cm bed of oomicrite and finally another 5 cm marl bed.

The Oolite Marl has long been recognised as a richly fossiliferous deposit notably from the occurrence of brachiopods (*Plectothyris fimbria, Pseudoglossothyris,* and rhynconellids) in the marl bands.

In this section the Oolite Marl is capped by some 1.8 m of Upper Freestone, a medium-grained oosparite.

From this point near the Devil's Chimney a footpath can be followed along the top of the quarry faces, across the ancient camp with its ditches, to a small quarry in the Ragstones at the eastern end of the top of the main quarry face (SO 949 185). Much of the surrounding hill top has been quarried away and now the quarry floors have reverted to farmland. The exposure described here is at the top of the very steep incline (formerly a tram line; Richardson 1904, p. 117) leading down to the old building in the main quarry.

The first beds to be examined occur below the small quarry floor and are exposed in the uppermost part of the main quarry. This is the top of the Upper Freestone – an oosparite with patches of micrite matrix. Some 5.5 m are exposed but the total thickness is about 9 m. It is of particular interest that this formation shows evidence of subaerial exposure prior to the deposition of the Lower Trigonia Grit which overlies it unconformably. The primary cement of the oosparite is non-ferroan but some of the ooliths have then been dissolved away, perhaps under subaerial conditions, and the resultant cavity infilled with ferroan calcite.

Between this and the floor of the small quarry are 2.7 m of Lower Trigonia Grit i.e. Middle Inferior Oolite. The basal 15 cm of the Lower Trigonia Grit is marly and was regarded by Buckman (1895) as the Harford sands equivalent. It is followed by a series of ferruginous biomicrites all of which have distinctive bioclasts of bivalve shell with a prismatic structure believed to be of the Pterinacea. One softer rubbly 90 cm bed (Fig. 5) contains chamosite ooliths. The hard band just below the quarry floor has an abundance of small serpulid tube clusters which can be observed in the field.

In the small quarry proper, some 3 m of rubbly, highly fossiliferous

(*Gryphaea sublobata*) sandy biomicrite make up the section. This is the Gryphite Grit of the Middle Inferior Oolite.

The detailed succession within the Ragstones (M. Inf. Ool.) of Leckhampton Hill was first described by Buckman (1893). This was followed by an abbreviated account in 1895. The exact spot on Leckhampton Hill which was used in the description is unknown; there is a large area of disused quarry workings on the hill top. It is difficult to recognise the details of Buckman's subdivisions in the small quarry here described. This is unfortunate as this area is the type locality for the Lower Trigonia Grit and the Buckmani Grit (see Donovan & Hemingway 1963). In particular the 3.5 m of sandy limestones of the Buckmani Grit cannot be recognised between the Lower Trigonia Grit and the Gryphite Grit. It is curious that no detailed section has been described subsequently, not even by Richardson (1904) who described almost every other available section in the Cheltenham district. This anomaly needs detailed palaeontological investigation before a final explanation can be offered. It is possible that the thicknesses recorded by Buckman were only of very local application, and that the two beds of hard ferruginous biomicrite with serpulids (75 cm and 23 cm thick), here included with the Lower Trigonia Grit, might represent a thinner lateral equivalent of the Buckmani Grit. Richardson (1904, p. 116) comments that the passage from the Buckmani Grit to the Gryphite Grit is transitional rather than abrupt thus supporting this hypothesis.

15
Westbury and northern Wiltshire

A. N. Insole and C. A. Wright

Objectives This itinerary provides an opportunity to study the scattered sections in the Middle Jurassic to Upper Cretaceous successions of northern Wiltshire.

Special Features Permission must be obtained in advance to visit the following localities:

Caen Hill, Devizes: Messrs. Rendell & Sons Ltd., Elmtree Court, Long Street, Devizes, Wilts. Devizes 2151.

Beggar's Knoll Quarry and Westbury Cement Works: Associated Portland Cement Manufacturers Ltd., Westbury Works, Trowbridge Road, Westbury, Wilts. Westbury 2481.

Maps

Ordnance Survey	One-Inch Sheets	157 (Swindon), 166 (Frome) 167 (Salisbury)
	1:50 000 Sheets	173 (Swindon & Devizes), 183 (Yeovil & Frome) 184 (Salisbury & The Plain)
	1:25 000 Sheets	ST 85, ST 86, ST 95, ST 96, ST 97
Geological Survey	One-Inch Sheets	265 (Bath), 266 (Marlborough) 281 (Frome), 282 (Devizes)

Main Reference Arkell (1933).

Outline Geology The region covered by this itinerary consists of Middle Jurassic to Upper Cretaceous rocks, mainly dipping gently east or south-east (Fig. 1). The stratigraphic succession is as follows:

	Thickness (metres)
CRETACEOUS	
Upper Chalk: Soft white chalk with some flint bands; hard limestone (Chalk Rock) at base	*c.*75
Middle Chalk: Hard massive white chalk with infrequent flints; hard, somewhat nodular limestone (Melbourne Rock) at base	*c.*27
Lower Chalk: Soft whitish chalk with occasional hard bands passing down into blocky beds with flint nodules; the base is marked by thin beds of sandy chalk and glauconitic marls with phosphatic nodules (Chloritic Marl)	*c.*75
Upper Greensand: Greenish-grey sands and sandstones with rare doggers; Malmstone at base	*c.*43
Gault: Clay; grey and sandy at the top becoming yellow, micaceous and silty at base	*c.*27
Unconformity	
Lower Greensand: Ferruginous sandstones and ironstones	*c.*12
Unconformity	

JURASSIC		
Portland Beds: Yellow and buff sands with irregular beds of calcareous sandstone		c.12
Kimmeridge Clay: Blue-brown clay with septaria		152
Corallian Beds: A variable sequence of sands, clays, limestones and ironstones. In the north of the region (Calne area), at the top of the formation, the Upper Calcareous Grit is composed of a ferruginous clay with some ironstone bands; to the south, the ironstones become dominant (Westbury Ironstone). Beneath the Upper Calcareous Grit is a laterally variable group of limestones (Coral Rag), including, at some horizons, coral rich beds (e.g. Steeple Ashton Coral Bed). At the base of the formation, the Lower Calcareous Grit consists of yellow sands and sandstones with subordinate clays		36
Oxford Clay:	Dark blue-grey clay	c.180
Kellaways Beds:	Blue-grey, sometimes sandy clay	c.6
Cornbrash:	Pale rubbly and marly limestones	c.5
Forest Marble:	Interbedded clays and thin shelly limestones	15–30
Bradford Clay:	Interbedded clays and thin shelly limestones	3
Great Oolite:	Oolitic limestones	

Although the outcrop pattern is basically aligned parallel to the regional NE strike, the distribution of rock types has been complicated by normal faulting and folding. The faults mainly trend ENE to WSW, and, although they only affect Jurassic rocks, they probably represent post-Cretaceous reactivation of Variscan structures in the basement. The Vale of Pewsey and Trowbridge Anticlines have ENE to WSW axial traces. The former is responsible for the marked easterly indentation of the Chalk outcrop and the isolated occurrence of Portland Beds south of Devizes. The Trowbridge Anticline is a less conspicuous structure but it is responsible for the narrow Cornbrash inlier running NE from Trowbridge.

Topographically the area divides into two distinct parts: the Jurassic lowlands to the north and north-west and the Chalk uplands to the south and east. The alternation of Jurassic limestones and clays combined with the low regional dip produces a series of low escarpments separated by broad, flat, often marshy tracts. Where there is an unconformable cover of Lower Greensand on the Upper Jurassic low rounded hills are produced. The eastern and southern boundaries of the area are marked by the twin Upper Greensand and Chalk escarpments, which rise up to the high Chalk downlands.

Itinerary

1 **Calne Car Park** (ST 997 709) In Calne follow the signposts to the public car park by municipal offices. The old quarry face at the back of the car park provides a section in the Calne Freestone, a local development within the Coral Rag (Corallian). Lithologically, the rock is a biomicrite, the bioclasts including abraded gastropods and shell fragments. Sedimentary features include interbedded thin clay laminae, trough bedding and the trace fossil *Ophiomorpha*.

2 **Caen Hill, Devizes** (ST 982 614) From Calne take the B 3102 and A 342 to Devizes. On entering Devizes bear sharp right along the A 361 toward Trowbridge. Just over 1 km along this road near the foot of the Upper Greensand escarpment as the dual carriageway ends, the entrance to the disused brickpit can be seen on the right (north) side of the A 361. The section visible is in the Lower Gault (*Hoplites dentatus* zone, Middle Albian) and was first recorded by Jukes-Browne (1900, p. 252):

3. Clay, blue grey .... 1.2 m
2. Clay, yellow-brown with ferruginous concretions .... 0.3 m

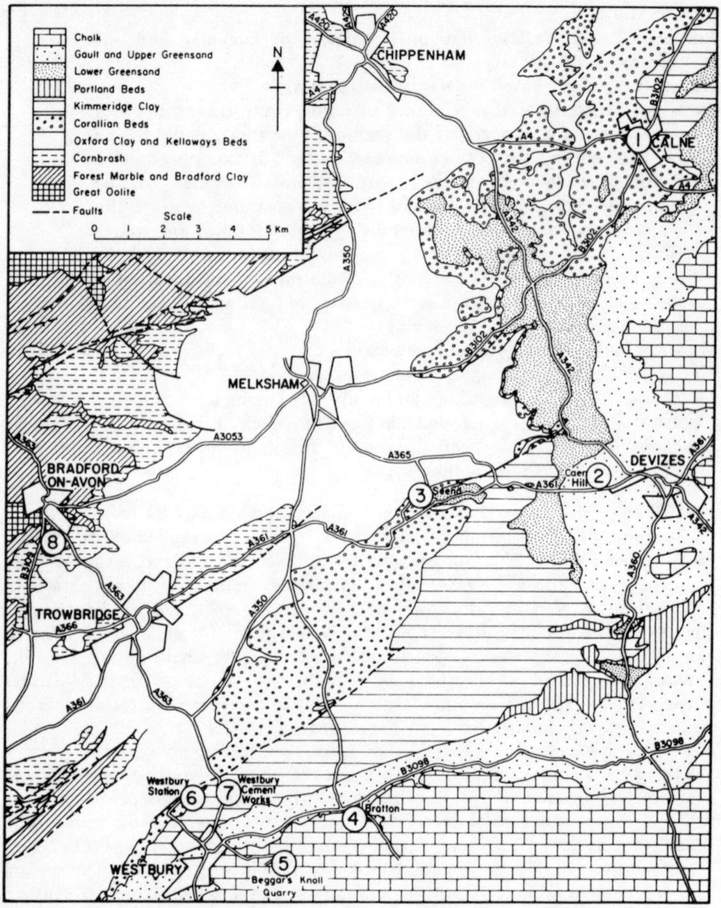

Fig. 1 Locality and geology map

1. Clay, micaceous, dark blue-grey with scattered septarian nodules . . . . 4.5 m
Base of Section

Fossils are rare but *Hoplites dentatus* and other *Hoplites* species have been found in beds 1 and 2.

3 Seend (ST 937 611) Continue along the A 361 towards Trowbridge until the village of Seend is reached. An outlier of Lower Greensand occurs around this village. These beds may be examined in sections along a private lane and in small quarries on the south side of the minor road opposite the Bell Inn leading to Seend Cleeve. These exposures show about 3.5 m of ferruginous sands and sandstones. Coarser pebbly beds and crossbedding are visible near the base of the sequence. Ironstone occurs in layers, frequently forming crusts around ovoid patches of sand. Earlier records (Cunnington 1850) show that the base of the sands rests unconformably on an eroded and bored surface of the Kimmeridge Clay. Although no fossils have been obtained recently, Cunnington obtained a fauna remarkable for the dominance of patellid gastropods (*Scurria, Aomaea, Loxotoma*). Casey (1961, p. 563), who gives the most recent published faunal list, places the Seend iron-sands in the *Parahoplites cunningtoni* subzone of the *Parahoplites nutfieldensis* zone (Upper Aptian). The sands were quarried intermittently between 1855 and 1874 for iron, smelting being carried out on the spot (Lamplugh, Wedd & Pringle 1920, p. 220).

4 **Bratton** (ST 917 522) Return to the A 361 and continue westwards to crossroads with the A 350. Turn left and proceed southwards, taking the minor road through Steeple Ashton and the signposted road to Bratton. On entering the village turn left opposite the fine mediaeval Court House, and continue for 1 km to crossroads. Proceed straight across the B 3098 and continue southwards on road to Patcombe Hill. At 400 m from the crossroads there is an Army range sign and opposite this is a small pit which exposes about 3 m of Upper Greensand of typical lithology with a layer of doggers near the top. A few metres north of the pit, a lane leads to the church, off which a footpath forks southwards up the Chalk escarpment. In isolated exposures along this path, the transition between the Upper Greensand and the Lower Chalk can be seen.

5 **Beggar's Knoll Quarry** (ST 892 508) From Bratton, proceed to Westbury along the B 3098, which runs along the Upper Greensand platform at the base of the major Chalk escarpment. At 0.5 km east of Westbury town centre, a narrow road through Newtown should be taken which climbs to northern slopes of Beggar's Knoll. At the top of the hill there is a crossroads and also the entrance to a very large quarry.

The section in this quarry exhibits greyish-white, occasionally nodular Chalk (Turonian) with two minor bands of flint. The macrofauna recorded (Waldman 1965) consists of serpulids (*Glomenula gordialis*), brachiopod fragments, *Ostrea* spp., *Inoceramus* spp., the ammonite *Collignoniceras woolgari*, teeth of pycnodontids and lamnids, fish scales and an incompletely preserved specimen of the coelacanth *Macropoma mantelli*.

6 **Westbury Station** (ST 866 525) Retrace the route to the B 3098 and proceed to Westbury railway station. Cross the railway to the north of the station and, after a short distance, take the turning at Ham Post Office signposted Trowbridge. Take the first turn to the right (Hawkeridge Road). After passing over the railway again, a rough track leads off to the left and does a U-turn to pass under the road via a tunnel. Immediately, on the other side of the tunnel, a section is visible in the Westbury Ironstone. The section is composed of approximately 3 m of brown and reddish-brown oolitic ironstone with concentrations of *Liostrea* at certain levels. This provides one of the few remaining visible sections in the Westbury Ironstone. This deposit was worked from 1856 up to about the turn of the century (Lamplugh, Wedd & Pringle 1920), as can be seen from the extensive workings in the area.

7 **Westbury Cement Works** (ST 880 527) Return to the centre of Westbury via the station and take the A 350 road towards Trowbridge. After approximately 1 km a road forks off to the right to a cement works. A pit has been excavated in the Kimmeridge Clay (*Aulacostephanus mutabilis* zone). The lithology is the typical blue-grey clay with septarian nodules and irregular masses of grey limestone. The fauna from the clays consists of abundant, compressed bivalves and ammonites.

8 **Bradford-on-Avon** (ST 826 601) Continue along the A 350 and A 363 to Bradford-on-Avon via Trowbridge. Before reaching the centre of the town, take the fork left along B 3109 until after 0.5 km the bridge over the Kennet and Avon Canal is reached. Immediately before crossing the bridge a track leads southwards along the east bank of the canal for 100 m to a disused quarry. The section here was originally described by Cox (1941, p. 33); the only available section is that given by Stinton & Torrens (1968, p. 246).

FOREST MARBLE

8. Clay, grey-brown .... 0.15 m
7. Limestone, shelly .... 0.15 m
6. Clay, grey .... 2.4 m
5. Limestones, flaggy, shelly separated by 0.15 m clay band .... 0.6 m

BRADFORD CLAY

4. Clay, grey .... 0.6 m
3. Limestone, platy, shelly and fine-grained, impersistent .... 0.3 m
2. Clay, grey .... 2.7 m

GREAT OOLITE

1. Oolitic limestone, shelly forming the floor of the pit

A small fault occurs across the pit so that bed 1 forms a higher platform at the west end of the exposure (i.e. nearest the canal). The rich fauna of the Bradford Clay, which includes the brachiopods *Digonella digona, Eudesia cardium, Dictyothyris coarctata* and *Epithyris bathonica,* is best obtained from bed 2. A solitary specimen of *Clydoniceras* was found *in situ* in bed 3 (Stinton & Torrens 1968). The Bradford Clay is assigned to the *Clydoniceras discus* zone (Bathonian).

16
The Quaternary of the North Somerset area

A. B. Hawkins

Objectives

Since 1970 it has been accepted that this area has been glaciated. This excursion visits many of the places where the evidence was obtained. Localities en route are given where periglacial deposits, abandoned beaches and modern estuarine sedimentation can be studied.

Special Features

Permission is required to visit the following locality:
The Holly Lane quarry: Mr. Moon (Clevedon 4451) or Mr. W. H. Seward, Lucki We Cottage, Holly Lane (Clevedon 2933).

Maps

Ordnance Survey	1:50 000 Sheets 172 (Bristol & Bath)
	182 (Weston-super-Mare)
	1:25 000 Sheets ST 35, ST 36, ST 45, ST 46, ST 47, ST 57
Geological Survey	One-Inch Sheets 264 (Bristol), 280 (Wells)
	1:25 000 Sheet ST 47

Main References

Hawkins & Kellaway 1971; Hawkins 1972.

Outline Geology

Although Trimmer (1853) described the presence of Pleistocene erratics in gravels at Norton's Wood (east of Clevedon) and in the Kenn and Yatton areas, his paper was largely ignored by subsequent workers. Harmer (1907) suggested some of the gorges were formed as overflow channels when ice blocked the Flax Bourton Gap, which he considered to be the early valley of the Bristol Avon. Again, his suggestion was generally ignored or discounted by later writers, the authors concluding that the anomalous drainage was the result of superimposition.

In 1969, a dark reddish-brown shelly material, resembling till, was examined in a manhole excavation at the entrance to Kenn Court Farm (loc. **10**). This included a pocket of yellow very fine sand which, when processed, was found to contain a rich microfauna. J. W. Murray noted the absence of any *Ammonia beccarii* (Linné) while more than half the microfauna consisted of an *Elphidium* species similar to *E. clavatum* Cushman; which is characteristic of cold brackish water environments. Following this recent find of something that could convincingly be argued was glacial, a continuous surveillance of trenches in the area was undertaken, mainly by Dr. D. D. Gilbertson. This established the lithological types present, the faunal changes and complex sedimentary history of the glaciogenic deposits of the Kenn area. These deposits, mapped by the Institute of Geological Sciences on Sheet 264 as Burtle Beds, are now known to be an accumulation of till and outwash gravels cut by interglacial streams and overlain around the periphery by marine sands and gravel (Gilbertson & Hawkins, in press).

In April 1970 it was realised that the excavation for the M 5 through the Failand Ridge at Court Hill, Clevedon (loc. **5**) had cut into a 25 m deep valley,

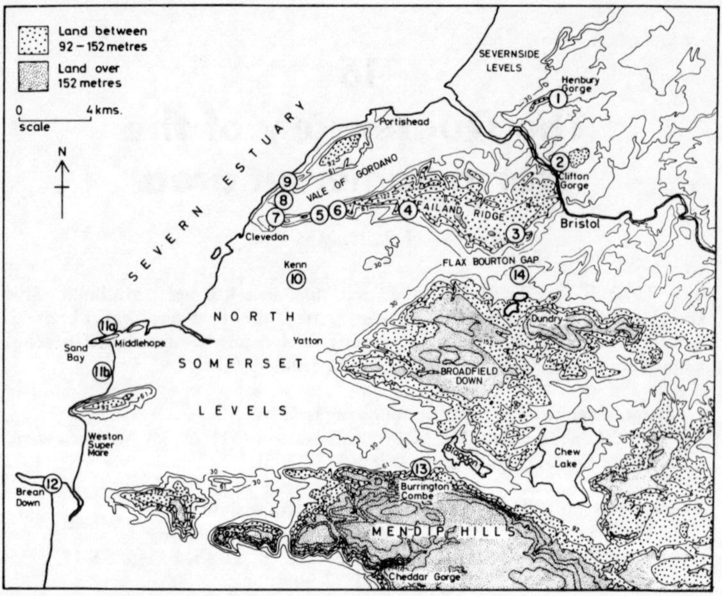

Fig. 1 Locality map

infilled with gravel, cobbles and boulders, mainly of Carboniferous Limestone plus some Pennant Sandstone, Devonian sandstone and conglomerate and a little Greensand chert. Incorporated within these 'gravels' were patches of till and pockets of sand. The approximately 16 m thick deposit of rudaceous material rested upon an impersistent layer of till beneath which was up to 8 m of loose current-bedded sand. In the 'gravels' many of the coarser clasts had been cemented at point contact while the gravel size material was often so calcreted it resembled a conglomerate.

Having located one valley breaching the hillside, an inspection of that part of the Failand Ridge indicated that the deposits mapped as Dolomitic Conglomerate and Keuper Marl in the Tickenham Valley (loc. 6), might also be of glacial origin. In September 1970, 13.7 m of deposits similar to those seen in the Court Hill Channel were proved in a borehole, see Fig. 5. A further borehole in Swiss Valley (the East Clevedon Gap, loc. 7) failed to confirm the presence of Triassic deposits but instead, proved Head overlying sands with some gravel, resting directly on Carboniferous Limestone.

A glacial origin is the most plausible explanation for the deposits at Kenn and in the channels at the western end of the Failand Ridge. As no significant pollen bearing deposit has been found, the dating of the glaciation is debatable. The freshwater deposits and the marine shelly sands around the perimeter are clearly older than the shell date of 33,750±1750 BP (I 4850), (Gilbertson & Hawkins, in press) hence they are probably of Ipswichian age. Whether the channels such as Swiss Valley, Court Hill, Tickenham Valley as well as the Clifton and Henbury Gorges are Wolstonian or Anglian in age cannot yet be established.

It is suggested that much of the area was covered by ice in the Anglian Glaciation, and that the channels were cut at this period. This earlier erosion of the channels and the diversion of the River Avon, would help to explain the quite high watershed in the Flax Bourton Gap. The more restricted Wolstonian

Glaciation advanced as far as the Kenn area, where typical till and outwash gravels accumulated. Erosion in the Late Wolstonian removed much of the glacial material, but in the Kenn area relics were left in 'basins' in the Keuper Marl. The rise in sea level during the Ipswichian allowed the erosion of abrasion platforms at Holly Lane (loc. **8**), Middle Hope (loc. **11a**) and Brean Down (loc. **12**) as well as the deposition of marine and freshwater sediments at Kenn. During the Devensian the sea level was low and most of the Ipswichian deposits were eroded. The cold wet and cold dry climate caused scree and aeolian sands to accumulate, as seen at Holly Lane and Brean Down. Approximately 9 000 years ago, during the Flandrian Transgression, the sea entered the deep valleys, resulting in the accumulation of the thick sands, silts and clays of the present day Levels (Hawkins 1971).

Itinerary

1 Trym Gorge at Henbury In order to appreciate the geomorphological significance of the Trym Gorge, it is advisable to view the Kings Weston Ridge from about 1.5 km to the north. A good vantage point is on the bridge which takes Wyck Beck Road over the railway line at ST 5703 7973 . From the bridge the incision cut by the Trym into the wooded Carboniferous Limestone rocks of the Kings Weston Ridge can be seen by looking SW. Because of the sinuous and wooded nature of the gorge, it is not possible to appreciate the full amplitude of the valley, see Fig. 2. Looking westwards, the minimum ground level is not much higher than the bridge height. From the spot height on the Ordnance Maps the col south of Haw Wood is about 41 m OD. The River Trym, however, instead of continuing a westwardly course across the Keuper Marl towards the Severnside Lowlands, takes a right-angle bend and cuts a 60 m deep incision into the Carboniferous Limestone, joining the River Avon at Sea Mills.

The size of the present River Trym can be approximated by noting the small northern tributary which runs along the west side of the bridge embankment and, by returning approximately 130 m south, where the 1 m wide eastern tributary has a thin alluvial surface veneer at about 37 m OD. The bend in the river at ST 567 796 can be seen by returning towards Bristol along Wyck Beck Road and after 400 m, turning right along Passage Road then left into Standfast Road. After travelling SW for about 60 m, turn right into Chamney's Avenue and proceed into Peverell Close. The right-angle bend in the river can be inspected in the field at the end of the road.

As the gorge is heavily wooded, it is not possible to appreciate the geomorphical features from the gorge itself. Hawkins (1972) has recorded that the Trym, after crossing the almost flat Triassic plain, follows the line of the Henbury Fault in the northern part of the Kings Weston Ridge but, over much of its course, there is no apparent geological control.

The present catchment of the Henbury Trym is less than 5 km^2; hence the relationship of the stream to the gorge is that of a misfit river.

No evidence has been found so far that ice entered the Hallen Gap. In some boreholes in the Avonmouth area it has been possible to identify thin till deposits beneath the estuarine alluvium, resting on the Keuper Marl. To the north, the gravel deposits under part of industrial Severnside have been interpreted as a glacial outwash deposit. The evidence, therefore, is strongly in favour of ice having advanced up the Severn to the west of Hallen Gap and it is presumed that at this period the blocked river took and created a new course.

2 The Clifton Gorge (Sea Walls) (ST 560 747) With a party the best vantage point is the Sea Walls area of Durdham Down. From the last locality return towards Bristol for 4 km and, having reached the open park, take the right fork at the roundabout into Parry's Lane. Within 100 m take the left-hand turning into Saville Road and then, at the crossroads, proceed straight across, reaching Sea Walls after 1.5 km.

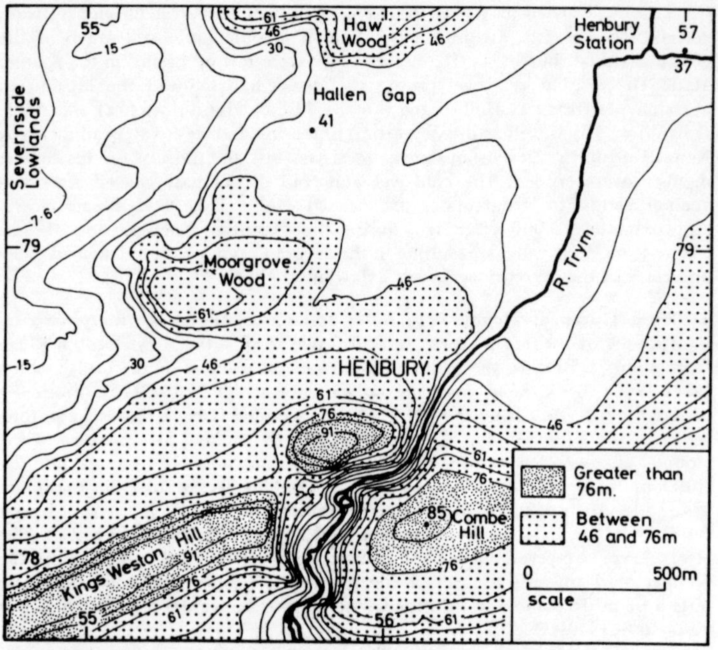

Fig. 2 Map of the Trym Gorge at Henbury (loc. 1)

Sea Walls is situated at the edge of the Durdham Downs plateau, at a height of about 85 m OD. The surface was probably eroded during the Rhaetic transgression, although there may have been some modification during Late Tertiary times (Hawkins & Kellaway 1971).

The rock gorge is about 97 m deep in the Sea Walls area, but the bedrock channel is infilled with up to 20 m of superficial deposits. Quarrying and later road and rail excavations have produced many artifically steep faces above HWST level, while in places rock ledges have been modified to facilitate navigation. Southwards, between Sea Walls and Hotwells the trend of the gorge, and any slight deviation from it, may have been controlled by the alignment of the Mesozoic and post-Mesozoic discontinuities (Hawkins & Kellaway 1971). The truncated Carboniferous Limestone (dipping c.30° at 140°) is seen at the top of the Great Quarry, 500 m SSE.

To the NW the gorge widens out and becomes less steep as it crosses the Old Red Sandstone. The alluvial lowlands and Severn Estuary are visible past the M 5 Avon Bridge.

Bradshaw (1966) summarised a large number of papers on the origin of the Clifton Gorge. Because of the outliers of Rhaetic and Liassic strata and the accepted geomorphological opinion at the time, he concluded 'The drainage is superimposed'. There is little doubt now, with the evidence that has accumulated since 1969, that the origin of the gorge is glacial; although whether it is an overflow channel or tunnel valley is not yet established.

3 **Providence** Southwards from Sea Walls, around the edge of The Downs, take the B 4468 down into the gorge. Although it is not possible to stop a vehicle, an excellent exposure of Dolomitic Conglomerate is visible 250 m down the Bridge Valley Road (ST 5640 7377). Here the breccia, with limestone clasts up to 1 m in size, is cut by the sub-vertical discontinuities

which probably control the gorge alignment. Joining The Portway, those travelling by car may park just before passing under the Suspension Bridge in order to look back at the eastern face of the gorge, where various thrust planes cut the Carboniferous Limestone.

Crossing Cumberland Basin by following the South West signs, the B 3128 should be taken towards Clevedon. To the north of the road is the location of the Ashton Park Borehole (ST 5633 7146), drilled in 1952–3. Here Kellaway (1967, p. 53) recorded 'One unexpected result followed the excavation of the mud tanks at the drill site. Here pebbles and fragments of honey coloured Greensand chert were found at a depth of 5½ ft (1.68 m) in soft red mud which had previously been thought to be undisturbed Keuper Marl'. Kellaway at that time decided this was a Head deposit; but the similarity to some of the Keuper-rich tills in the Kenn area and the evidence now available, suggests this may have been a thin till relic. How much of the Head marked on the geological maps is in fact glacial material is not yet known, due to lack of exposures.

The lower part of the possible glacial spillway shown in Fig. 3 runs through Ashton Park. However, from the quite sharp bend to the left near one of the lodges, the road follows the spillway for 750 m. At the first crossroads those travelling by coach should leave their transport and take the road to the south, towards the village of Providence. Where the houses begin, 300 m south of the junction, a short walk along the footpath to the right (marked Birdwell) will allow the visitor to see the open, lower terrain to the west. The narrow eastern footpath ascends the gentle slope to the field which contains the head of the spillway, at about 115 m OD. From here the dry valley follows a sinuous course towards the southern part of the Clifton Gorge. The presence of this spillway, with no catchment, could be cited as evidence that ice once blocked the Flax Bourton Gap, probably the original valley of the Bristol Avon. Although no exposures have been available recently in the gap, a sewer trench near the river south of Gable Farm (ST 503 707) did show patches of till overlying Keuper Marl in areas labelled as Head on Geological Sheet 264. The quarry near the Providence crossroads also contains deposits of Greensand chert and quartz in a silty clay, found in solution hollows and pipes, which may be redeposited glacial material (Colborne *et al.* 1974).

Returning to the coach it is worth taking the northern footpath to examine an exposure of Clifton Down Limestone. Here the topmost beds display excellent evidence of borings (Hawkins & Kellaway 1971).

4 Moat House Farm (ST 487 727) From Providence follow the B 3128 along the top of the Failand Ridge. Temporary exposures in the area south of Moat House Farm indicated over 4 m of silty clay, with much manganese staining and some gravel and cobbles of Greensand chert and iron nodules. Whilst this exposure may in part represent decomposed Black Rock Dolomite (Colborne *et al.* 1974), a similar deposit exceeding 2.9 m found at ST 491 729 on Black Rock Limestone indicates the material is not confined to one lithological type. The presence of these drift deposits, at the head of the steep sided Caswell Valley, suggests future work will prove that the top of the Failand Ridge has been glaciated.

5 Court Hill Approximately 3.5 km after joining the B 3130 road turn right into the Hill Lane, indicated on a signpost as Cadbury Camp. Those travelling by coach must walk from here. Nearly 1 km up the lane, past where the tarmac road ends, is the path leading to a bridle bridge over the motorway. There is little exposure now, but calcreted gravel is still visible in a small cutting on the approach to the bridge and in a few scattered patches in the motorway excavation. Where the clasts are small, they are calcreted and resemble a conglomerate, but where the clasts are larger, the cementation in only at point contact. Although interpreted during the site investigation as

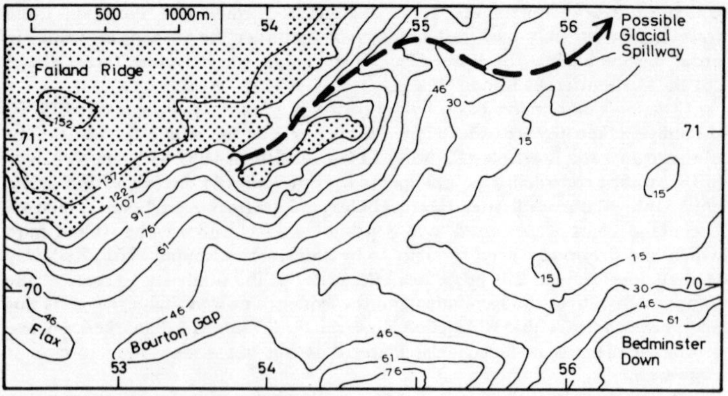

Fig. 3 Glacial spillway, Providence (loc. 3)

Dolomitic Conglomerate, these gravels were found on exposure to contain Greensand chert.

The edges of the Pleistocene channel can be approximated on the west side, at the soil/rock junction just north of the bridge. On the east a sloping rock/soil interface about 50 m north of the bridge again indicates the change from Carboniferous Limestone to channel fill material. 75 m south of the bridge on the same side, the western edge of the channel can just be determined where scattered rock exposures occur.

The first record of these gravels was by Trimmer (1853). In Reynolds (1916b) they are shown only as non-Carboniferous rocks, while the present geological maps do not indicate the deposits at all. This channel, from a col height of 74 m OD, to a bedrock depth of 50 m OD, is infilled with approximately 16 m of 'gravel' with some sand pockets and a little till, overlying a current-bedded uncemented sand. Frequently, there is a thin (less than 1 m) veneer of till overlying the basal sands. In the lower half of the rudaceous deposit the gravel, cobble and boulder sized clasts generally occurred as layers dipping up to 37° north.

The lowlands north of the col are known as the Vale of Gordano. Parts of this area have a present surface level of only 4.5 m OD compared with the height of the estuarine alluvium near the coast, which averages over 7 m. Consequently in much of the wet 'backlands' peat has accumulated. Jefferies *et al.* (1968) have described a sand bar beneath the surface running across the Vale in the vicinity of Weston Drove (Fig. 4). Augering behind this bar they found peat of Zone III age. To the south, the slope to the alluvial lowlands is now known to be covered by up to 8 m of sandy and gravelly Head (Hawkins & Kellaway 1971). Thus the horticultural use of this slope will be influenced by the sandy soils, as well as the southern aspect.

6 Tickenham Valley From Court Hill, proceed eastwards along the rough trackway into the next col (ST 448 724). Here the trackway runs along a ridge at the southern end of the valley, (Fig. 5). As the lane is on the ridge and because of the wooded slopes, the U-shaped nature of the valley does not appear as clear in the field as on the contoured map.

With a careful search on the pathway south of the col, a few lumps of conglomerates can still be seen. Because of the similarity between this and the 'gravel' in the Court Hill channel, a borehole was located in the col and drilled to 13.7 m. The hole, at ST 4485 7242, proved material very similar to the Court Hill deposits and again the microfauna in the sands were dominantly of Jurassic and Cretaceous-age. This, plus the rare presence of Greensand chert,

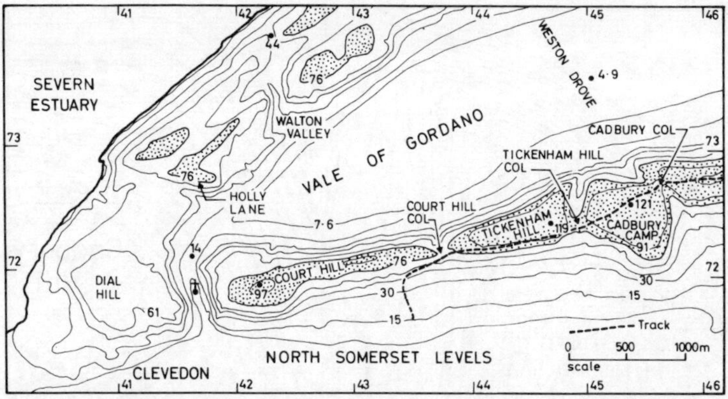

Fig. 4 Vale of Gordano and Tickenham Hill (loc. 5, 6, 7, 8 & 9) heights in metres

proved conclusively that the material was not Dolomitic Conglomerate, as indicated on Geological Sheet ST 47. Unfortunately the borehole stopped at 75 m OD and did not prove the contact between the 'gravels' and the Carboniferous Limestone. The break of slope in the valley, plus a subsequent geophysical survey, suggest that the channel base may be 70–73 m OD.

A careful search of the area marked Keuper Marl on the maps, yields fragments of Greensand chert. Degraded pits in this material expose coarse, brown, aeolian sand becoming more clay rich towards the surface. Thus the sandy deposits in the bottom of the valley used for horticulture are Head, rather than Keuper Marl as shown on the maps. At the northern end of the channel, excavation for the M 5 Wynhol Viaduct proved extensive solution hollows in the Carboniferous Limestone, now infilled with loose sand.

Although in the field and in Fig. 4 one of the most pronounced breaks in the hill crest is east of Cadbury Camp, this incision follows the line of a fault and, therefore, is more likely to have been originally a Triassic feature.

From the Tickenham Col parties may either return westwards along the track or take the northern path (Fig. 5) and, having passed under the M 5 Wynhol Viaduct, turn left along the road at the northern foothill. From here it is approximately 3.5 km to Swiss Valley.

7 Swiss Valley (East Clevedon Gap) This valley breaches the Carboniferous Limestone ridge and effectively joins the Vale of Gordano to the North Somerset Levels. The floor of the gap averages 8 m OD, rising to a watershed of 14 m OD in the north (ST 416 721). The sides rise for 51 m, in places precipitously.

There is no geological evidence to imply the valley follows a fault. On Geological Sheet ST 47 the valley is shown floored with Keuper Marl. Recent sewer trenches, however, failed to confirm the presence of Keuper Marl. For this reason, a borehole was drilled in 1970 just south of the church (ST 4164 7166). This borehole, at 7.5 m OD, proved 13 m of sand with some gravel, overlying Carboniferous Limestone (Hawkins 1972). The lack of Trias in this borehole means that, at this point at least, there is now no evidence for a Keuper Marl infill in the valley, hence the feature being of Triassic age. Greenly (1919) described his search for Triassic rock in the valley, and concluded that the very small patches of calcreted gravels that can be found, with some difficulty, behind the cottages at ST 4176 7172 were probably Dolomitic Conglomerate. The finding of two pieces of Greensand chert plus the similarity to the Court Hill deposits suggest they are more likely to be of

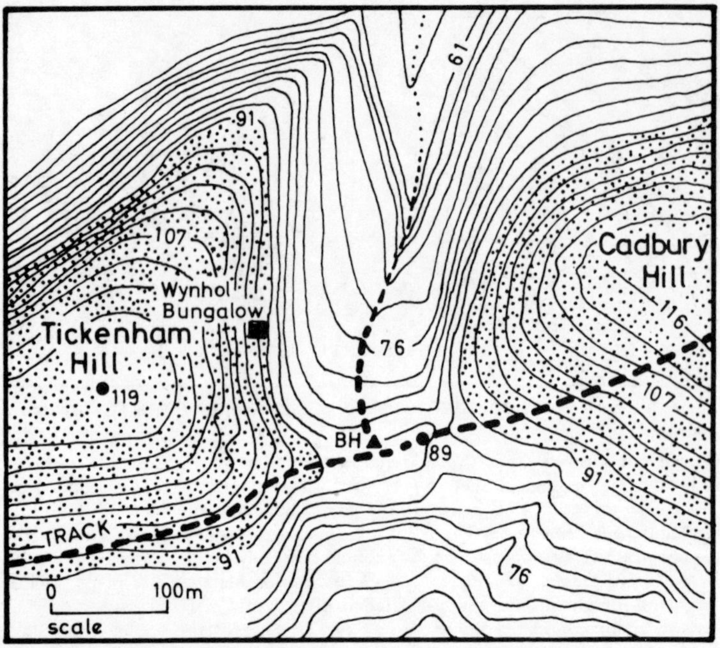

Fig. 5 Tickenham Hill (loc. 6) heights in metres

Quaternary age.

The borehole by the church proved the bedrock at this locality to be at -5.5 m OD. It is unlikely that the single borehole in the valley is in the deepest part of the channel. Hence, it is reasonable to assume the rock valley may have a depth of -10 m OD. It is significant that the bedrock level in the Hotwells area of the Clifton Gorge, which contains a large tidal river, is -12 m OD (Hawkins 1962). The wide, -12 to -15 m OD valley at the outlet of the Vale of Gordano could not be explained by Hawkins in 1967. While a meltwater origin would satisfactorily explain the size of the feature, its maximum known depth of less than -15 m OD does impose a probable depth to the Swiss Valley channel, if both were formed by water being released from ice occupying the North Somerset Levels.

8 Holly Lane (ST 418 727) Leaving Swiss Valley on the B 3124 (Portishead) road turn into Holly Lane, the first road on the left. The entrance to the exposure is just past the sports field. A request to enter via the gates just past Lucki We Cottage, should be made to Mr. Moon of the tyre depot (Clevedon 4451) or Mr. Seward who lives in the cottage (Clevedon 2933). To reach the exposure it is necessary to walk through the old tyre dump and NE through the nettles and brambles alongside the fence. Large parties are advised not to visit this exposure because of difficulty with access.

The exposure has recently been described by Gilbertson & Hawkins (1974), while significant previous papers on the area are by Davies (1907), Hinton (1907), Reynolds (1907a), Greenly (1922), Palmer & Hinton (1929) and Palmer (1934). The present exposure shows a section through the periglacial Head deposits, (Fig. 6). The Head is made up of three sandy horizons within a sandy breccia with varying clast sizes. The Head has an apparent dip of 30° and rests against a Carboniferous Limestone cliff.

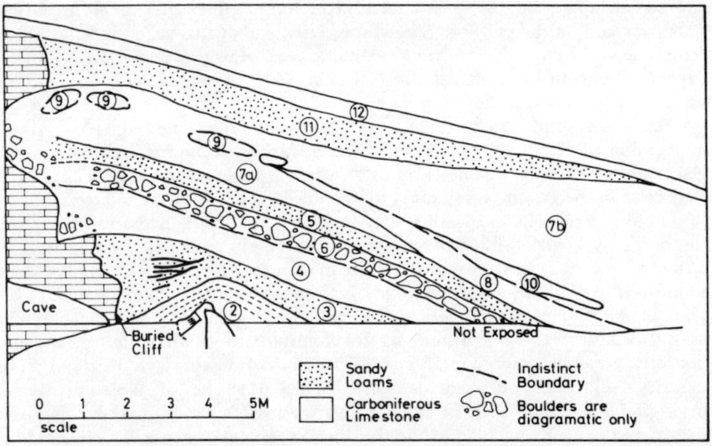

Fig. 6 Section at Holly Lane (loc. 8) explanation in text

The most pronounced layers are Units 5 and 6 of Gilbertson & Hawkins (1974). The main boulder horizon, Unit 5, contains clasts up to 0.5 m in length, in complete contrast to the silty sand above. The sands contain several molluscan species with *Hygromia hispida* (Linnē) and *Pupilla muscarum* (Linnē) being the most common. If collecting is done, care should be taken to obtain fresh samples as some species that have been collected in the past are obviously contaminants. The foraminiferid assemblage from the sand horizons suggest a life habitat of fine muddy sands of near marine environment, while the presence of *Cibicides lobatulus* (Walker & Jacob) indicates firm substrate such as weeds or animals. The sands contain abundant fragments of comminuted shells, yet the foraminiferid assemblage, which probably lived in the Bristol Channel, does not show abrasion, suggesting the microfossils may have been carried into the area by ice, before being transported to their present position by wind. The Head deposit, therefore, is considered to be a periglacial breccia which formed in a cold wet climate with a sandy matrix and sand layers which accumulated as an aeolian deposit in cold dry periods.

While the quarry was being worked there was a limestone slope to the SW of the present face which showed a clear abrasion notch (see photo in Palmer & Hinton 1929). In winter, when the vegetation is less, this can be seen approximately 15 m to the SW of the Head exposure (see Fig. 108 in Gilbertson & Hawkins 1974). The notch has a breccia infilling, similar to that of the main section. In the previous literature the notch was recorded as 15 m. Recent levelling however, has shown the abrasion features to be at about 20 m OD, i.e. about 15 m above the alluvial lowlands.

As there is no biostratigraphical evidence, the dating of the Head and the underlying erosional features has been decided mainly by dating back from the present. As there are no soil horizons in the Head, it is considered to be of Devensian age. Thus the underlying abrasion features could be Ipswichian or older. It has been suggested by Gilbertson & Hawkins that they may represent the maximum sea level of the Ipswichian interglacial.

9 Walton Valley From Holly Lane return to the B 3124 and proceed for 1 km towards Portishead, before turning left in the village of Walton-in-Gordano. It is not easy to appreciate the shape of the valley through the Black Rock Limestone while ascending the road. The Lower Limestone Shale Group has been selectively eroded and the outcrop now forms the hollow occupied by the golf course. The top of the valley at 44 m OD is narrower and on the Old Red Sandstone. The head of this valley is only 200 m from the sea, hence there is no catchment area. Hawkins & Kellaway (1971) suggested this to be a glacial overflow channel, associated with ice banking against the Portishead-Clevedon ridge.

10 Kenn and Yatton Leaving Clevedon by the B 3133, travel approximately 2 km to the village of Kenn. Although the first evidence of the glaciation of Somerset came from here, there is little to be seen now except a slight rise in the ground level, most noticeable just south of the village.

On the geological maps this area is shown as Burtle Beds, following the discovery by Welch (1956) of marine shells when he was mapping the area. In 1969 however, at the entrance to Kenn Court Farm (ST 415 688) a dark reddish-brown shelly material closely resembling till was located. This contained a pocket of yellow silty sand which had a microfauna indicative of a cold brackish water environment. It is now known that the glaciogenic deposits are extremely complex. Gilbertson & Hawkins (in press, A) have described till and outwash deposits, believed to be of Wolstonian age; inter-glacial stream and peripheral marine sands and gravelly deposits considered to be Ipswichian, while the whole complex is generally veneered by coversands of Devensian age. Except in the highest areas the glaciogenic deposits are covered by Flandrian silts and clays.

Passing through Kenn the next village is Yatton. Rutter (1829) and Trimmer (1853) both recorded Pleistocene deposits from here, while Greenly (1919), referring to the deposits exposed at Yatton station, writes, 'the formation recalls the true boulder clays, but the extreme rarity of striated stones, the feebleness of the striations, and the almost total absence of erratics, forbid us to regard it as such'. Hawkins (1972) examining temporary exposures considered it to be largely till or redeposited till which, with a few breaks, could be followed in the sewer trenches from the deposits at Kenn.

11a Middle Hope Continuing southwards along the B 3133, take the right turning at the junction with the A 370 towards Weston-super-Mare. After entering the town, turn right and follow the signposts to Sand Bay. Travelling by coach it is easiest to continue along Kewstoke Road to the bay. Cars and minibuses can be taken to the northern end of the road (ST 330 659), but coaches should be left in the bus area approximately half way along the seafront (ST 333 646).

Go through the gateway in the lane leading to the café, just past the entrance to the car park, and immediately inside the gateway take the path on the left which goes steeply up the hillside. At the top, negotiate the stile and bear left across the grass towards a stone wall. Keep to the north of the wall and follow it as it runs parallel to the coast, right down to the shore at Swallow Cliff (ST 3245 6605). Although referred to by many authors since Sanders (1841), until recently there has been no comprehensive description – see Gilbertson & Hawkins (in press,B).

At low tide it is possible to see a wide sea-weed covered abrasion platform along the northern coast of Middle Hope. It is quite possible that this has a similar age to the Howe Rock Platform of ApSimon *et al.* (1961) and Donovan (1962); hence being Middle Devensian. In the area where the volcanic ash is exposed (see Excursion 4.2, loc. 1.1) selective weathering has produced a small discordant bay. In this the modern beach pebbles have formed a storm beach which is probably higher than would occur on an open coast, due to the westerly aspect.

The Swallow Cliff abandoned beach is situated against a fossil cliff, with the notch and abrasion platform being approximately 5 m above the modern storm beach. The fossil cliff is approximately 3 m high. Banked against it are a series of silty sands and breccia overlain by the main abandoned beach (Unit 8,

Unit	Description	Approximate Thickness
9	Angular gravel to boulder sized clasts of Carboniferous Limestone in sandy silt matrix.	0.4 m
8	Rounded gravel to boulder sized clasts (up to 0.45 m long including one with borings) with a sandy, very shelly matrix. The shells identified by Dr. Gilbertson include *Macoma balthica* (Linnē), *Littorina littoralis* (Linnē), *L. littorea* (Linnē), *L. saxatilis* (Olivi), *Trophonopsis truncatus* (Strom), *Nucella lapillus* (Linnē), *Ocenebra erinacea* (Linnē), *Nassarius reticulatus* (Linnē), *Lora* sp., *Ostrea* sp., *Cardium* sp. (probably *edule*) fragments.	1.1 m
7	Sandy layer with some gastropod and bivalve shells, especially *Macoma balthica.*	0.20 m
6	Sub-angular to sub-rounded gravel and cobble sized fragments including one vein quartz pebble.	0.15 m
5	Sandy layer with some sandrock nodules and some gravel sized clasts.	0.18 m
4	Sub-angular gravel to boulder sized clasts of Carboniferous Limestone and lava. Occasional gastropods in sandy matrix.	0.25 m
3	Impersistent sandy silt layer.	0.1 m
2	Sub-angular to sub-rounded gravel and cobble sized clasts of Carboniferous Limestone in silty sand matrix.	0.1–0.3 m
1	Abrasion platform at c. 12.5 m OD.	

Table 1

Table I). The lower sequence is that of a solifluction breccia interbedded with aeolian sands, not dissimilar to that at Holly Lane. The sands contain quite abundant warm water microfauna. The breccia consists entirely of locally derived material. The abandoned beach is unusual in having such a high shell content, both whole and comminuted. A list of the fauna in the beach layer is given in Table I.

The age of the beach in uncertain, but most likely Ipswichian. Callow & Hassall (1969) published two radiometric dates for shell samples, but the Middle Devensian age probably represents a minimum age, this being at the maximum range of the radiocarbon technique. If the beach is Ipswichian the Head overlying it would be Devensian. The age of the deposits, Units 2–7, beneath the beach are not so definite. It could be argued that they are typical of cold wet/cold dry climates and therefore, Wolstonian in age, implying the abrasion features are Hoxnian. Alternatively, it is possible that the abrasion platform is Ipswichian overlain by aeolian and hillwash material before being resubmerged, resulting in the accumulation of the beach material.

The level of the abandoned beach is variously referred to in the literature as the 10 ft (3 m) or 45 ft (15 m) feature. In an area such as the Severn Estuary where there is an extreme tidal regime, it is much better to give the height relative to the modern beach feature. Therefore, it is suggested that abandoned beach should, in future, only be referred to as the 5 m beach, namely 5 m above the modern storm beach and about 12.5 m OD.

Returning along the wall, an overgrown deposit of sand, probably aeolian,

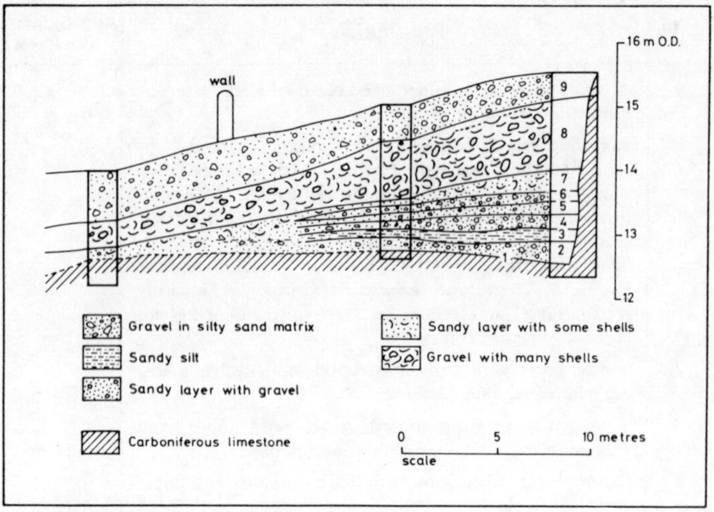

Fig. 7 Swallow Cliff, Middle Hope (loc. 11a)

can be located by the animal burrows. It is not known how deep these sands are, but again they contain an abundant microfauna.

Macfadyen (1970) makes reference to another abandoned beach in North Bay (ST 339 665). Here only patches of calcreted gravel remain near the top of the southern side of a 'ravine'. Its height is again about 15 m OD and hence similar to that at Swallow Cliff. Again sands, probably of aeolian origin, can be seen in the old animal burrows.

11b Sand Bay (ST 33 65) Those interested in modern estuarine deposits should take time to examine the sedimentation in Sand Bay, but extreme care is required. Near LWST level the sediment at the surface changes to resemble a liquid mud.

In the north, the spread of the *Spartina* marsh has caused concern in recent years. The atypical appearance is a result of the local authority's attempt to restrict its growth. Sections in the marsh exhibit the layered nature of the clay, silt and occasional sand grade material. Through the sediment, the rotted *Spartina* roots have left sub-vertical pipes, 2–4 mm in diameter. Along the seaward edge of the marsh, the ebbing waters have produced typical runnels, approximately 0.7 m deep, 0.5 m wide and 1–2 m apart. The depth of the runnel depends on the height of the cliff created by the waves at HWNT level.

To the south, there is a small beach. Seaward of this beach, sand is the normal deposit in areas where the beach is convex. However, in the hollows clay and silt grade material frequently accumulate. Sometimes during periods of erosion, minor inversions of relief are caused when the more cohesive muddy sediments form the higher areas with small (0.1 m) cliffs forming around the seaward margin. The small cliffs will be vertical or stepped, depending upon whether the material is homogeneous or laminated. Careful examination of the sediment shows the bioturbation taking place. In these muddy sediments the most common beach fauna are *Macomabalthica* (Linnē) and *Corophium arenarium* Crawford. Either by examining the small cliffs or by digging where there are small petaloid patterns on the surface sediment, the bivalves can be seen in their vertical growth position, approximately 70 mm down. Further out, the sandy areas are generally characterised by the mounds

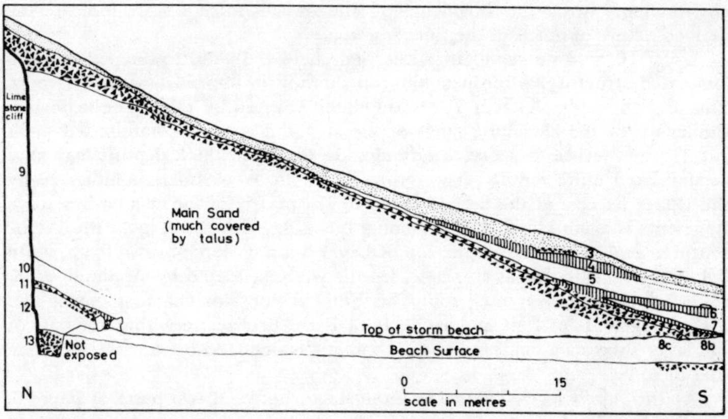

Fig. 8 Section at Brean Down (loc. 12) explanation in text (after ApSimon *et al.* 1961)

and siphon cones of the lug worm *Arenicola* marina (Linnē). The U-shaped *Arenicola* burrows may penetrate 0.3 m beneath the surface and, because of this and the sandy sediment, it is less easy to appreciate their shape than the 50 mm deep *Corophium* burrows.

Approximately half way down the 1–2 km wide beach, the surface is again runnelled. On the landward side the runnels are only 0.2 m deep. Whilst the sediment in the ridges is laminated, that in the floor of the runnel frequently consists of a lag sediment of sand, shells and occasionally randomly orientated fallen blocks, eroded from the intervening ridges. Although the regular runnel pattern is caused by the main ebb flow across a particular sediment type, the final ebb waters follow meandering dendritic 'runnels', independent of the regular ones. The main fauna in this zone is the small gastropod *Hydrobia ulvae* (Pennant) although, especially in the upper part, *Macoma balthica* are again quite common. A modern fauna list can be obtained in Boyden & Little (1973).

12 **Brean Down** (ST 295 587) From Middle Hope and Sand Bay the visitor should map read his own journey to Brean Down. The main sand cliff described by ApSimon *et al.* (1961) is at HWST level on the south side of the Carboniferous Limestone promontory at its junction with the sand dunes. At present it is not possible to see much of Layer 9, Fig. 8, but the other layers of ApSimon *et al.* can be distinguished. Below is a brief description of the deposits; for a full description reference should be made to the detailed text in their paper.

The Lower Breccia (Layer 13) is just visible at present. It consists of a loose angular Head, with boulders up to 1 m and is reported to contain vole, arctic fox and reindeer, suggesting the environment in which it accumulated was cold and moist. Layer 12 is a clayey, silty sand with occasional angular limestone fragments. Some reindeer antler and rodent remains have been recovered from this layer. The smaller proportion of rock fragments plus a matrix, which is partly derived by hillwash and partly aeolian, imply the climate may well have been drier, with occasional wetter periods. The Middle Breccia and Bone Bed (Layer 11) begins with a limestone breccia generally of gravel sized clasts in a silty sand matrix. The bone bed proper is reported by ApSimon *et al.* to contain lemming, varying hare, arctic fox, elephant, horse, reindeer and bird bones, with horse and reindeer being dominant. The presence of the non-marine shells *Pupilla muscorum* (Linnē) suggests a cold environment, while the fragments of *Macoma balthica* simply indicate the derived nature of much of the finer fraction.

Layer 10 is more sandy than the beds below. In the lower part it has a laminated structure, while near the top there is an impersistent thin breccia. The aeolian sands of Layer 9 are now much covered by talus. The deposit is similar to the modern dune sands of the area and contain comminuted shells and foraminiferids. In a dry, windy climate the 15 m thick deposit may have accumulated quite rapidly. The return to a cold, wet climate is indicated by the Upper Breccia at the base of Layer 8. The matrix of the angular limestone fragments is again sandy, but becoming more clay rich towards the top. At the southern end of the section the top of Layer 8 is a dark red-brown loam, up to 0.4 m thick. This almost stoneless deposit was considered by ApSimon *et al.* to be a mature *terra fusca* soil. This horizon contains charcoal, bones and shells, fragments of flint and artefacts of Early Bronze Age. Thus the base of the layer suggests a cold, moist environment which later became warmer than that of today.

At the beach level, Layer 7 is intercalated between two parts of Layer 6. The Beaker Sand (Layer 7) is probably of aeolian origin and accumulated quickly. It contains animal bones and molluscan shells. Layer 6 is a sandy, silty clay with some sub-rounded limestone cobbles. Animal remains include sheep, goat and possibly wild boar. The local accumulation of large boulders of weathered limestone have been interpreted as the remains of man-built structures. Most of Layer 5 is sand, probably again of aeolian origin. Over these are the sandy loams with some weathered cobbles of limestone (Layer 4). Archaeological study of this layer has proved the presence of Iron Age occupation. Above this, the deposits are dominantly sands (Layer 3), which have probably been accumulating steadily since Iron Age times.

ApSimon *et al.* have suggested that the breccia/sand sequence accumulated in Late Devensian-Flandrian times.

At the western end of the promontory the bevelled reef feature of the Howe Rock Platform occurs between 0 and 5 m OD. Because the top of the abrasion platform is below HWST level it is considered not to be a modern feature but probably of Middle Devensian age; although this would represent a higher sea level than would be accepted by many authorities. Above this pronounced lower abrasion platform there are indications of benches at 12–14 m OD and also at 20 m OD.

13 **Rickford Channel** (ST 493 592) From Brean Down to Rickford the journey can be made by several routes. Unfortunately, due to the irresponsible behaviour of some parties, the gravel deposit at 82 m OD on Bleadon Hill (Findlay *et al.* 1972) is no longer accessible. Visitors unfamiliar with the area, may wish to travel via Cheddar Gorge and Burrington Combe. The relationship between the glaciation of the area and the evolution of these gorges has not yet been worked out; although the gravel in the fans at the mouth of the gorges (Findlay 1965) was undoubtedly transported by streams swollen by meltwater. Whether this was from snowfields, isolated ice caps or from larger ice sheets has not yet been determined.

The quickest route from Brean Down is via the A 38 and A 368. This road passes through the 40–50 m deep Rickford Valley, following the line of the glacial spillway. This meandering, dry valley is incised in the Dolomitic Conglomerate, which, in this vicinity forms a spur on the northern flank of the Mendip Hills.

From the entrance to Combe Lodge, the spillway valley slopes at about 1:50 to the west and approximately 1:20 to the NW, towards the River Yeo (see Fig. 9). Thus the valley has no catchment and is left 'hanging' 60 m up the side of the Mendip Hills. Hawkins & Kellaway (1971) suggest it was formed when the Yeo Valley was blocked by ice and developed as an ice-marginal

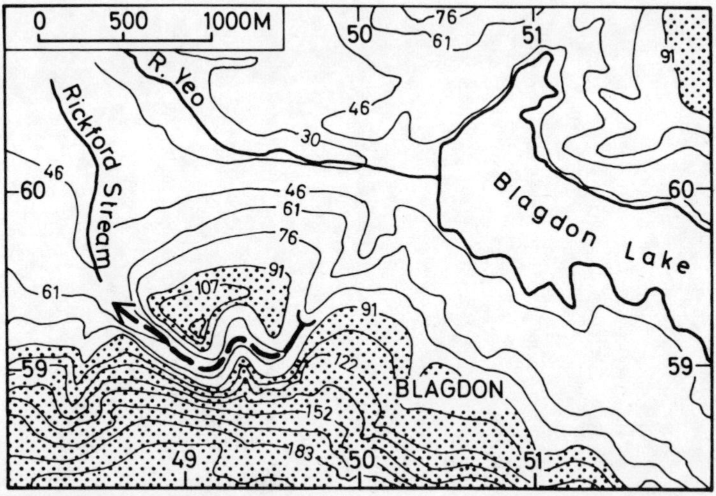

Fig. 9 Rickford Channel (loc. 13)

feature.

14 Bedminster Down (ST 573 701) Continuing along the A 368, between Ubley and Compton Martin the low watershed between the westward draining Yeo valley and the eastward draining Chew is visible. Hawkins (1972) has suggested that the Chew, like many other rivers north of Eastern Mendips, was diverted when ice blocked the previous seaward outlet.

From Compton Martin take the left hand turning towards Chew Stoke, passing Chew Valley Reservoir on your right. Leaving the village of Chew Stoke take a left hand turning towards Bristol. After 1.5 km continue straight across at the road junction and follow this road over Dundry Hill into Bristol.

On reaching the traffic lights at the junction with the A 38, turn left and, after about 200 m, park in order to view the city from the vantage points on the north side of the main road. This is an excellent area from which to see the way that Bristol developed in the lowlands of the River Avon and spread on to the surrounding hills. The wide lowlands continue through the Flax Bourton Gap to the west, but the River Avon makes a sudden change in direction and passes through the Clifton Gorge. With the evidence available now, there is little doubt that this channel has a glacial origin.

Bibliography

ACKERMANN, K. J. & CAVE, R. 1968. Superficial deposits and structures including landslip, in the Stroud District, Gloucestershire. *Proc. Geol. Ass.*, **78**, 567–86, 4 figs.

AGER, D. V. 1954. The genus *Gibbirhynchia* in the British Domerian. *Proc. Geol. Ass.*, **65**, 25–51, 7 figs., 1 pl.

1956. Field Meeting in the central Cotswolds. *Proc. Geol. Ass.*, **66**, 356–365.

AGER, D. V., DONOVAN, D. T., KENNEDY, W. J., McKERROW, W. S., MUDGE, D. C. & SELLWOOD, B. W. 1973. The Cotswold Hills. *Geol. Assoc. Guide* No. 36 (2nd Edition). 34 pp., 4 figs.

AMBROSE, T. & ROMANO, M. 1972. New Upper Carboniferous Chelicerata (Arthropoda) from Somerset, England. *Palaeontology*, **15**, 569–578, 2 figs., 2 pls.

ApSIMON, A. M., DONOVAN, D. T. & TAYLOR, H. 1961. The Stratigraphy and Archaeology of the Late-Glacial and Post-Glacial Deposits at Brean Down, Somerset. *Proc. Univ. Bristol Spelaeol. Soc.*, **9**, 67–136, 22 figs., 6 pls.

ARKELL, W. J. 1933. *The Jurassic System in Great Britain.* xii + 681 pp., 97 figs., 41 pls. Oxford: Clarendon Press.

ARKELL, W. J. & DONOVAN, D. T. 1952. The Fuller's Earth of the Cotswolds and its relation to the Great Oolite. *Jl geol. Soc. Lond.*, **107**, 227–253, 2 figs., 4 pls.

ATKINSON, T. C. 1971. The dangers of pollution of limestone aquifers, with special reference to the Mendip Hills. *Proc. Univ. Bristol Spelaeol. Soc.*, **12**, 281–90, 2 figs., 1 pl.

ATKINSON, T. C., DREW, D. P. & HIGH, C. 1967. Mendip Karst Hydrology Project, Phases 1 and 2. *Occ. Publ. Wessex Cave Club*, Series 2, (1), 38 pp., 17 figs., 8 pls., 7 tbls.

BAILY, W. H. 1864. On the occurrence of fish-remains in the Old Red Sandstone of Portishead, near Bristol. *Geol. Mag.*, **1**, 293.

BARRINGTON, N. & STANTON, W. I. 1972. *The complete caves of Mendip.* (2nd Edition). Cheddar Valley Press, Cheddar, Somerset. 155 pp., 9 pls., 5 maps.

BASSETT, D. A. & BASSETT, M. G. (eds.) 1971. *Geological Excursions in South Wales and The Forest of Dean.* 267 pp., 82 figs. Cardiff: Geol. Assoc. S. Wales Group.

BLYTH, F. G. H. 1952. Malvern Tectonics – A Contribution. *Geol. Mag.*, **89**, 185–194, 3 figs.

BLYTH, F. G. H. & LAMBERT, R. St. J. 1970. Chemical data from the Malvernian of the Malvern Hills, Herefordshire. *Jl geol. Soc. Lond.*, **125** (for 1969), 543–555, 1 fig.

BOULTER, M. 1968. A Species of Compressed Lycopod Sporophyll from the Upper Coal Measures of Somerset. *Palaeontology*, **11**, 445–457, 4 figs., 2 pls.

BOULTON, W. S. 1904. On the igneous rocks at Spring Cove, near Weston-super-Mare. *Jl geol. Soc. Lond.*, **60**, 158–168, 8 figs.

BOYDEN, C. R. & LITTLE, C. 1973. Faunal Distribution in Soft Sediments of the Severn Estuary. *Est. Coast Mar. Sci.*, **1**, 203–223, 8 figs., 2 tbls.

BRADSHAW, R. 1966. The Avon Gorge. *Proc. Bristol Nat. Soc.*, **31**, 203–220, 9 figs., 1 pl.

BRADSHAW, R. & HAMILTON, D. 1967. Conjugate gypsum veins at Blue Anchor Point, Somerset. *Proc. Bristol Nat. Soc.*, **31**, 305–309, 5 figs., 1 pl.

BROOKS, M. 1968. The geological results of gravity and magnetic surveys in the Malvern Hills and adjacent districts. *Geol. J.*, **6**, 13–30, 5 figs., 2 pls.

1969. Reconnaissance seismic refraction surveys in the Malvern Hills area. *Geol. J.*, **6**, 147–160, 7 figs.

1970. Pre-Llandovery tectonism and the Malvern structure. *Proc. Geol. Ass.*, **81**, 249–268, 7 figs.

BROOKS, M. & DRUCE, E. V. 1965. A Llandovery conglomeratic limestone in Gullet Quarry, Malvern Hills, and its conodont fauna. *Geol. Mag.*, **102**, 370–381, 1 fig., 1 pl.

BUCHANAN, R. A. & COSSONS, N. 1969. *Industrial Archaeology of the Bristol Region.* 335 pp., 20 figs., 32 pls. Newton Abbot: David & Charles.

BUCKLAND, W. & CONYBEARE, W. D. 1824. Observations on the South-western Coal District of England. *Trans. geol. Soc. Lond.*, Ser.2, **1**, 210–316, 2 figs.

BUCKMAN, S. S. 1893. The Bajocian of the Sherborne District: its relation to subjacent and superjacent strata. *Q. Jl geol. Soc. Lond.*, **49**, 479–522, 2 figs.

1895. The Bajocian of the Mid-Cotteswolds. *Q. Jl geol. Soc. Lond.*, **51**, 388–462, 8 figs., 1 pl.

1901. Bajocian and contiguous deposits in the North Cotteswolds: the main hill-mass. *Q. Jl geol. Soc. Lond.*, **57**, 126–55, 5 figs., 1 pl.

1887–1907. A monograph of the ammonites of the 'Inferior Oolite Series'. *Palaeontogr. Soc. (Monogr.)* **1**, 1–456, 104 pls.

1889. On the Cotteswold, Midford, and Yeovil Sands, and the division between Lias and Oolite. *Q. Jl geol. Soc. Lond.*, **45**, 440–74, 3 figs.

1897. Deposits of Bajocian age in the Northern Cotteswolds: the Cleeve Hill Plateau. *Q. Jl geol. Soc. Lond.*, **53**, 607–29, 6 diagrams, 1 map, 1 pl.

BULLEY, J. A. 1952. To Mendip for Coal – a Study of the Somerset Coalfield before 1830. Part 1. Output, Marketing and Techniques of Mining. *Proc. Som. Arch. nat. Hist. Soc.*, **97**, 46–78.

BUTCHER, N. E. 1962. The tectonic structure of the Malvern Hills. *Proc. Geol. Ass.*, **73**, 103–123, 1 fig., 2 pls.

BUTLER, M. 1972. *Conodont faunas and stratigraphy of certain Tournaisian sections in the Bristol-Mendip area.* Unpublished Ph.D. Thesis, University of Bristol.

BUTLER, WILLIAMS, B. P. J. & BRADSHAW, R. 1972. A new exposure of the Old Red Sandstone – Lower Limestone Shale transition at Portishead, Somerset. *Proc. Bristol Nat. Soc.*, **32**, 151–155, 2 figs.

CALLOW, W. J. & HASSALL, G. I. 1969. National Physical Laboratory Radiocarbon Measurements. VI. N.P.L. *Radiocarbon* **11**, 130–136.

CASEY, R. 1961. The stratigraphical palaeontology of the Lower Greensand. *Palaeontology*, **3**, 487–621, 14 figs., 8 pls.

CHALONER, W. G. & COLLINSON, M. E. 1975. An Illustrated Key to the commoner British Upper Carboniferous plant compression fossils. *Proc. Geol. Ass.*, **86**, 1–44, 15 figs., 1 tbl.

CHANNON, P. J. 1950. New and enlarged Jurassic sections in the Cotswolds. *Proc. Geol. Ass.*, **61**, 242–60, 1 fig., 1 pl.

CHATWIN, C. P. 1960. *British Regional Geology: The Hampshire Basin and adjoining areas.* (3rd Edition). iv + 99 pp., 42 figs., 7 pls. London: HMSO.

CLEW, K. R. 1970. *The Somersetshire Coal Canal and Railways.* 176 pp.,

15 figs., 30 pls. Newton Abbot: David & Charles.

COLBORNE, G. J., GILBERTSON, D. D. & HAWKINS, A. B. 1974. Temporary Drift Exposures on the Failand Ridge. *Proc. Bristol Nat. Soc.*, 33, 91–97, 2 figs., 1 pl.

COLEMAN, A. M. & BALCHIN, W. G. V. 1959. The origin and development of surface depressions in the Mendip Hills. *Proc. Geol. Ass.*, **70**, 291–309, 3 figs., 2 pls.

CONYBEARE, W. D. & PHILLIPS, W. 1822. *Outlines of the geology of England and Wales.* Part 1. 470 pp., map, sections. London: Wm. Phillips.

COWIE, J. W., RUSHTON, A. W. A. & STUBBLEFIELD, C. J. 1972. A correlation of Cambrian rocks in the British Isles. *Geol. Soc. Lond. Spec. Rep.*, **2**, 42 pp., 6 figs., 5 pls., 2 tbls.

COX, L. R. 1941. Easter Field Meeting, 1940. Bath. *Proc. Geol. Ass.*, **52**, 16–35, 3 pls.

1942. New Light on William Smith and His Work. *Proc. Yorks. geol. Soc.*, 25, 1–99, 6 figs., 8 pls.

CROOKALL, R. 1925a. On the Fossil Flora of the Bristol and Somerset Coalfield. Part 1. *Geol. Mag.*, **62**, 145–180, 4 pls.

1925b. On the Fossil Flora of the Bristol and Somerset Coalfield. Part 2. *Geol. Mag.*, **62**, 385–410, 3 pls., 3 tbls.

1929. *Coal Measure Plants.* 80 pp., frontisp., 39 pls. London: Ed. Arnold.

1955–1970. Fossil Plants of the Carboniferous Rocks of Great Britain. *Mem. geol. Surv. Gt. Br. Palaeont.*, **4**, pts. 1–6, 839 pp., 242 figs., 159 pls.

CUNNINGTON, W. 1850. On a section of the Lower Greensand near Devizes. *Q. Jl geol. Soc. Lond.*, **6**, 453–4.

CURTIS, M. L. K., DONOVAN, D. T., KELLAWAY, G. A. & WELCH, F. B. A. 1955. Geology *In* MacInnes, C. M. & Whittard, W. F. pp. 3–33, 5 figs., 1 pl., q.v.

DAVIES, D. K. 1969. Shelf Sedimentation: An example from the Jurassic of Britain. *J. sediment. Petrol.*, 39, No. 4, 1344–1370, 25 figs.

DAVIS, H. N. 1907. Supplementary Notes on the Clevedon Bone Cave and Gravels. *Proc. Bristol Nat. Soc.*, **1**, 188–189, 1 fig.

DAWKINS, W. B. 1864. On the Rhaetic Beds and White Lias of western and central Somerset; and on the discovery of a new fossil mammal in the grey marlstone beneath the Bone Bed. *Q. Jl geol. Soc. Lond.*, **20**, 396–412, 3 figs., 1 pl.

de la BECHE, H. T. 1846. On the formation of the rocks of South Wales and South Western England. *Mem. geol. Surv. Gt. Br. Palaeont.*, **1**, 1–296, 45 figs., 2 pls.

DONOVAN, D. T. 1955. Mesozoic *In* MacInnes & Whittard. pp. 23–28, q.v.

1956. The Zonal Stratigraphy of the Blue Lias around Keynsham, Somerset. *Proc. Geol. Ass.*, **66**, 182–212, 7 figs.

1958a. Easter Field Meeting: The Lower and Middle Jurassic Rocks of the Bristol District. *Proc. Geol. Ass.*, **69**, 130–140.

1958b. The Lower Lias section at Cannard's Grave, Shepton Mallet, Somerset. *Proc. Bristol Nat. Soc.*, **29**, 393–398, 1 fig.

1962. Sea Levels of the Last Glaciation. *Bull. geol. Soc. Am.* **73**, 1297–8, 1 fig.

1964. *In* McKerrow, W. S., Ager, D. V. & Donovan, D. T. Geology of the Cotswold Hills. *Geol. Assoc. Guide,* No. 36: 26 pp., 6 figs.

1969. Geomorphology and hydrology of the central Mendips. *Proc. Univ. Bristol Spelaeol. Soc.*, **12**, 63–74, 1 fig., 3 pls.

DONOVAN, D. T. & HEMINGWAY, J. E. 1963. *Lexique Stratigraphique International,* **1**, Fasc. 3aX Jurassique, 394 pp.

DOWN, C. G. & WARRINGTON, A. J. 1971. *The History of the Somerset Coalfield.* 283 pp., 58 figs., 28 pls. Newton Abbot: David & Charles.

DREGHORN, W. 1967. *Geology explained in the Severn Vale and Cotswolds.* Newton Abbot: David & Charles.

DREW, D. P., NEWSON, M. D. & SMITH, D. I. 1968. Mendip Karst Hydrology Project, Phase 3. *Wessex Cave Club, Occ. Publ.,* Series 2, No. 2, 28 pp., 6 figs., 7 pls., 5 tbls., 10 diagrams.

EARP, J. R. & HAINS, B. A. 1971. *British Regional Geology: The Welsh Borderland* (3rd Edition). xi + 118 pp., 46 figs., 11 pls. London: HMSO.

EDMONDS, E. A., McKEOWN, M. C. & WILLIAMS, M. 1975. *British Regional Geology: South-West England* (4th Edition), xi + 136 pp., 33 figs., 12 pls., 3 tbls. London: HMSO.

EYLES, J. M. 1969. William Smith: Some Aspects of His Life and Work. pp. 142–158, 1 fig. *In* Schneer, C. J. (ed.) *Toward a History of Geology.* Cambridge, Mass. and London: M.I.T. Press.

1974. William Smith's home near Bath; Tucking Mill. *J. Soc. Biblphy nat. Hist.* 7, 29–34, 1 fig., 1 pl.

FALCON, N. L. 1947. Major clues in the tectonic history of the Malverns. *Geol. Mag.,* **84**, 229–240, 1 fig.

FINDLAY, D. C. 1965. The soils of the Mendip district of Somerset. *Mem. Soil Surv. Gt. Br.* (Sheets 279 and 280). viii + 204 pp., 15 figs., 10 pls., 19 tbls., map.

FINDLAY, HAWKINS, A. B., & LLOYD, C. R. 1972. A Gravel Deposit on Bleadon Hill, Mendip, Somerset. *Proc. Univ. Bristol Spelaeol. Soc.* **13**, 83–87, 1 fig., 1 pl.

FOLK, R. L. 1959. Practical petrographic classification of limestones. *Bull. Am. Ass. Petrol. Geol.* **43**, 1–38, 7 figs., 5 pls.

1962. Spectral subdivision of limestone types. *Mem. Am. Ass. Petrol. Geol.,* **1**, 62–84, 7 figs., 1 pl.

1968. *Petrology of sedimentary rocks.* 170 pp. Austin, Texas: Hemphill's.

FORD, D. C. & STANTON, W. I. 1968. The geomorphology of south-central Mendip Hills. *Proc. Geol. Ass.,* **79**, 401–27, 9 figs.

GAYER, R. A. & STEAD, J. T. G. 1971. The Forest of Dean Coal and Iron-Ore Fields. *In* Bassett, D. A. & M. G. pp. 20–36, 5 figs., q.v.

GEIKIE, A. & STRAHAN, A. 1898. Volcanic group in the Carboniferous Limestone of North Somerset. *Summ. Progr. geol. Surv. Lond.,* 104–111, 1 fig.

GILBERTSON, D. D. & HAWKINS, A. B. 1974. Upper Pleistocene Deposits and Landforms at Holly Lane, Clevedon, Somerset (ST 419 727). *Proc. Univ. Bristol Spelaeol. Soc.* **13**, 349–360, 4 figs.

(in press A). The Pleistocene Succession at Kenn, Somerset. *Bull. geol. Surv. Gt. Br.* 37 figs., 2 pls., 8 tbls.

(in press B). The Pleistocene Deposits and Landforms at Swallow Cliff, Middlehope, Weston-super-Mare. *Proc. Geol. Ass.*

GILBERTSON, D. D. & MOTTERSHEAD, D. N. 1975. The Quaternary deposits at Doniford, west Somerset. *Fld Stud.* **4**, 117–129, 6 figs., 5 tbls., 3 pls.

GOTHAN, W. & REMY, R. 1957. *Steinkohlenpflanzen.* 248 pp., 221 figs., 6 tbls. Essen: Verlag Glückauf GmbH.

GREEN, G. W. 1958. The central Mendip lead-zinc orefield. *Bull. geol. Surv. Gt. Br.,* **14**, 70–90, 3 figs., 1 pl.

GREEN, G. W. & DONOVAN, D. T. 1969. The Great Oolite of the Bath Area. *Bull. geol. Surv. Gt. Br.,* No. **30**, 1–63, 6 figs., 6 pls.

GREEN, G. W. & WELCH, F. B. A. 1965. Geology of the country around Wells and Cheddar. *Mem. geol. Surv. Gt. Br.* (Sheet 280), x + 225 pp., 19 figs., 5 pls.

GREENLY, E. 1919. The Pleistocene Formations of Claverham and Yatton. *Proc. Bristol Nat. Soc.,* 5, 145–147.

1922. An Aeolian Deposit at Clevedon. *Geol. Mag.,* **59,** 365–376, 4 figs., 414–421.

GROOM, T. T. 1910. The Malvern and Abberley Hills. *Geol. Ass., Jubilee Volume,* 698–738, 6 figs., 1 pl.

HAHN, G. & HAHN, R. 1973. Viséan trilobites from Holwell, Somerset. *Palaeontology,* **16,** 551–561, 3 figs., 1 pl.

HALLAM, A. 1964. The origin of the limestone-shale rhythm in the Blue Lias of England: a composite theory. *J. Geol.,* 72, 157–169, 6 figs.

HALLAM, A. & SELWOOD, B. W. 1968. Origin of Fuller's Earth in the Mesozoic of southern England. *Nature, Lond.,* **220,** 1193–95.

HAMILTON, D. 1961. Algal Growth in the Rhaetic Cotham Marble of Southern England. *Palaeontology,* **4,** 324–333, 3 figs., 3 pls.

1962. Some notes on the Rhaetic sediments of the Filton By-pass Substitute, near Bristol. *Proc. Bristol Nat. Soc.,* **30,** 279–285, 2 figs., 1 pl.

1966. Disseminated galena in Rhaetic shales at Almondsbury, near Bristol. *Proc. Bristol Nat. Soc.,* **31,** 221–224.

HANCOCK, P. L. 1966. The stratigraphy and structure of the Inferior Oolite rocks of the Cotswolds. *Mercian Geol.* **1,** 275–281, 2 figs.

1969. Jointing in the Jurassic Limestones of the Cotswold Hills. *Proc. Geol. Ass.,* **80,** 219–41, 8 figs.

HANWELL, J. D. & NEWSON, M. D. 1970. The great storm and floods of July 1968 on Mendip. *Wessex Cave Club, Occ. Publ. Series* I, No. 2, 72 pp., 18 figs., 12 pls., 44 tbls.

HARMER, F. W. 1907. On the Origin of Certain Canyon-like Valleys associated with Lake-like Areas of Depression. *Q. Jl geol. Soc. Lond.,* **63,** 470–514, 4 figs., 5 pls.

HARRIS, T. M. 1938. *The British Rhaetic Flora.* 84 pp., 26 figs., 5 pls. Brit. Mus. (Nat. Hist.)., London.

HAWKINS, A. B. 1962. The Buried Channel of the Bristol Avon. *Geol. Mag.,* **99,** 369–374, 3 figs.

1966. The Geology of the Keynsham Bypass. *Proc. Bristol Nat. Soc.,* **31,** 195–202, 2 figs.

1967. The Geology of the Portbury Area. *Proc. Bristol Nat. Soc.,* **31,** 421–428, 3 figs.

1971. Sea Level changes around South West England. *Proc. Symp. Colston Res. Soc.* **23,** 67–88, 5 figs.

1972. Some Gorges of the Bristol District. *Proc. Bristol Nat. Soc.,* **32,** 167–185, 6 figs.

1973. The geology and slopes of the Bristol region. *Q. Jl. Engng. Geol.,* **6,** 185–205, 13 figs., 3 tbls.

HAWKINS, A. B. & KELLAWAY, G. A. 1971. Field Meeting at Bristol and Bath with Special Reference to New Evidence of Glaciation. *Proc. Geol. Ass.,* **82,** 267–291, 5 figs.

HEPWORTH, J. V. & STRIDE, A. H. 1950. A sequence from the Old Red Sandstone to Lower Carboniferous, near Burrington, Somerset. *Proc. Bristol Nat. Soc.,* **28,** 135–138.

HIGHLEY, D. E. 1972. Fuller's Earth. *Mineral Dossier* No. 3. *Mineral Resources Consultative Committee.* 26 pp., 5 figs., 2 tbls. London: HMSO.

HINTON, M. A. C. 1907. Note of the Occurrence of the Alpine Vole (*Microtus nivalis*) in the Clevedon Bone Deposit. *Proc. Bristol Nat. Soc.,* **1,** 190–191.

HOLLINGWORTH, S. E., TAYLOR, J. H. & KELLAWAY, G. A. 1944. Large-scale Superficial Structures in the Northampton Ironstone Field. *Q. Jl geol. Soc. Lond.,* **100,** 1–44, 14 figs., 3 pls.

ISON, W. 1948. *The Georgian Buildings of Bath from 1700–1830.* 211 pp., 136 pls. London: Faber & Faber.

JEFFERIES, R. L., WILLIS, A. J. & YEMM, E. W. 1968. The Late and Post-Glacial History of the Gordano Valley, North Somerset. *New Phytol.*, **67**, 335–348, 9 figs.

JONES, R. K., BROOKS, M., BASSETT, M. G., AUSTIN, R. L. & ALDRIDGE, R. J. 1969. An Upper Llandovery limestone overlying Hollybush Sandstone (Cambrian) in Hollybush Quarry, Malvern Hills. *Geol. Mag.*, **106**, 457–469, 3 figs.

JUKES-BROWNE, A. J. 1892. *The Building of the British Isles.* (2nd Edition). xi + 465 pp., 18 figs., 15 pls. London: Geo. Bell & Sons.

1900. The Cretaceous Rocks of Britain. I. The Gault and Upper Greensand of England. *Mem. geol. Surv. U.K.*, **1**, 1–499, 85 figs., 5 pls.

KELLAWAY, G. A. 1967. The Geological Survey Ashton Park Borehole and its bearing on the geology of the Bristol District. *Bull. geol. Surv. Gt. Br.*, No. **27**, 49–153, 8 figs., 6 pls.

1970. The Upper Coal Measures of South West England compared with those of South Wales and the Southern Midlands. *C. R. 6^e Congr. Intern. Strat. Geol. Carbonif. Sheffield*, 1967, **3**, 1039–1055, 6 figs., 1 pl.

KELLAWAY, G. A. & TAYLOR, J. H. 1968. The influence of Landslipping on the development of the City of Bath. *Rep. 23 Int. geol. Congr., Prague,* **12**, 65–76, 6 figs.

KELLAWAY, G. A. & WELCH, F. B. A. 1948. *British Regional Geology: Bristol and Gloucester district.* iv + 99 pp., 24 figs., 12 pls. London HMSO.

1955. The Upper Old Red Sandstone and Lower Carboniferous rocks of Bristol and the Mendips compared with those of Chepstow and the Forest of Dean. *Bull. geol. Surv. Gt. Br.*, **9**, 1–21, 2 figs., 1 pl.

KELLING, G. 1968. Patterns of sedimentation in Rhondda Beds of South Wales. *Bull. Am. Ass. Petrol. Geol.*, **52**, No. 12, 2369–2386, 11 figs.

KIDSTON, R. 1887. On the Fossil Flora of the Radstock Series of the Somerset and Bristol Coalfield (Upper Coal Measures). *Trans. R. Soc. Edinb.*, **33**, 335–417, 6 figs., 5 pls.

1923–5. Fossil Plants of the Carboniferous Rocks of Great Britain. *Mem. geol. Surv. Gt. Br. Palaeont.*, **2**, 1–681, 99 figs., 153 pls.

KLEIN, G. de V. 1965. Dynamic significance of primary structures in the Middle Jurassic Great Oolite Series, Southern England. *Spec. Publs Soc. econ. Palaeont. Miner. Tulsa* **12**, 173–191, 19 figs.

KÜHNE, W. G. 1956. The Liassic therapsid *Oligokyphus.* x + 149 pp., 66 figs., 12 pls. *Brit. Mus. nat. Hist.*, London.

LAMBERT, R. St. J. & REX, D. C. 1966. Isotopic ages of minerals from the Precambrian complex of the Malverns. *Nature, Lond.*, **209**, 605–606.

LAMBERT, R. St. J. & HOLLAND, J. G. 1971. The petrography and chemistry of the igneous complex of the Malvern Hills, England. *Proc. Geol. Ass.*, **82**, 323–351, 1 fig., 3 pls.

LAMPLUGH, G. W., WEDD, C. B. & PRINGLE, J. 1920. Special reports on the Mineral Resources of Great Britain. XII. Iron Ores (contd) – Bedded Ores of the Lias, Oolites and Later Formations in England. *Mem. geol. Surv. U.K.* iv + 1–240, 12 figs., 8 pls.

LIVINGSTONE, J. E. 1913 (for 1912). List of Publications bearing on the Geology of the Bristol District for the years 1875–1895. *Proc. Bristol Nat. Soc.*, 4th Ser. **3**, 119–133.

1914 (for 1913). List of Publications bearing on the Geology of the Bristol District for the years 1896–1913. *Proc. Bristol Nat. Soc.*, 4th Ser. **4**, 82–98.

1929 (for 1928). List of Publications bearing on the Geology of the Bristol District for the years 1914–28. *Proc. Bristol Nat. Soc.*, 4th Ser. **7**, 58–70.

LLOYD, A. J., SAVAGE, R. J. G., STRIDE, A. H. & DONOVAN, D. T. 1973. The geology of the Bristol Channel floor. *Phil. Trans. R. Soc.*, A. **274**,

595–626, 11 figs., 3 pls.

LONDSALE, W. 1835. On the Oolitic District of Bath. *Trans. geol. Soc. Lond.*, 3, 241–276, 1 fig.

MACFADYEN, W. A. 1970. *Geological highlights of the West Country.* London: Butterworths. 296 pp., 26 figs.

MacINNES, C. M. & WHITTARD, W. F. (eds.) 1955. *Bristol and its adjoining counties.* xiii + 335 pp., 24 figs., 37 pls. Bristol: Brit. Assoc. Adv. Sci.

McKERROW, W. S., AGER, D. V. & DONOVAN, D. T. 1964. Geology of the Cotswold Hills. *Geol. Assoc. Guide* No. 36, 26 pp., 6 figs.

MARTYN, S. 1876. On fish remains in the Bristol Old Red Sandstone. *Proc. Bristol Nat. Soc.*, 1, 141–144.

MATTHEWS, S. C., BUTLER, M. & SADLER, P. M. 1973. Lower Carboniferous successions in north Somerset; report by directors of field meeting. *Proc. Geol. Ass.*, **84**, 175–179.

MOORE, C. 1867. On Abnormal Conditions of Secondary Deposits when connected with the Somersetshire and South Wales Coal-basin. *Q. Jl geol. Soc. Lond.*, 23, 449–568, 6 figs., 4 pls.

MOORE, L. R. & TRUEMAN, A. E. 1937. The Coal Measures of Bristol and Somerset. *Q. Jl geol. Soc. Lond.*, 93, 195–240, 15 figs.

1942. The Bristol and Somerset Coalfields with Particular Reference to the Prospects of Future Development. *Proc. S. Wales Inst. Engrs.* 57, 180–305, 17 figs., 1 pl.

MORGAN, C. L. & REYNOLDS, S. H. 1904. The igneous rocks associated with the Carboniferous Limestone of the Bristol district. *Q. Jl geol. Soc. Lond.*, 60, 137–157, 4 figs., 2 pls.

MOSSOP, G. D. & SHEARMAN, D. J. 1973. Origins of secondary gypsum rocks. *Trans. Inst. Min. Metall.* 82, B147–B154, 9 figs., 2 tbls.

MURCHINSON, R. I. 1834. *Outline of the geology of the neighbourhood of Cheltenham.* (1st Edition). 40 pp., section. London: J. Murray.

MURRAY, J. W. 1969. The Inferior Oolite of the Cotswold scarp, Wotton-under-Edge to Leckhampton. *Proc. Bristol Nat. Soc.*, 31, 535–49, 5 figs.

NEWSON, M. D. 1972. Rickford and Langford resurgences, Mendip Hills, Somerset. A problem in limestone hydrology. *Proc. Univ. Bristol Spelaeol. Soc.*, 13, 105–112, 3 figs.

NORRIS, D. K. 1958. Structural Conditions in Canadian Coal Mines. *Geol. Surv. Bull. Can.*, 44, 1–54, 12 figs., 2 pls.

OWEN, T. R. 1971. The structural evolution of the Bristol Channel. *Proc. geol. Soc. Lond.*, No. 1664, 289–93, 1 fig.

PALMER, C. P. 1972. The Lower Lias (Lower Jurassic) between Watchet and Lilstock in North Somerset (United Kingdom). Newsl. *Stratigr.* 2, 1–30, 6 figs.

PALMER, L. S. & HINTON, M. A. C. 1929. Some Gravel Deposits at Walton, near Clevedon. *Proc. Univ. Bristol Spelaeol. Soc.*, 3, 154–161, 2 pls.

PALMER, L. S. 1934. Some Pleistocene breccias near the Severn Estuary. *Proc. Geol. Ass.*, 45, 145–161, 1 fig.

PALMER, T. J. & FÜRSICH, F. T. 1974. The ecology of a Middle Jurassic hardground and crevice fauna. *Palaeontology*, 17, 507–524, 3 figs., 3 pls., 1 tbl.

PARSONS, C. F. 1975. Ammonites from the Doulting Conglomerate Bed (Upper Bajocian, Jurassic) of Somerset. *Palaeontology*, 18, 191–205, 7 figs., 1 pl.

PENN, J. S. W., FRENCH, J., WHITTEN, D. G. A. & VINNICOMBE, J. 1971. The Malvern Hills. *Geol. Assoc. Guide*, No. 4, 36 pp., 16 figs.

PHILLIPS, J. 1844. Memoirs of William Smith, LL.D. ix + 150 pp., 18 pls. London: John Murray.

PHIPPS, C. B. & REEVE, F. A. E. 1964. The Pre-Cambrian–Palaeozoic boundary of the Malverns. *Geol. Mag.*, **101**, 397–408.

1967. Stratigraphy and geological history of the Malvern, Abberley and Ledbury Hills. *Geol. J.*, **5**, 339–368, 7 figs.

1969. Structural geology of the Malvern, Abberley and Ledbury Hills. *Q. Jl geol. Soc. Lond.*, **125**, 1–34, 4 figs., 3 pls.

PICK, M. C. 1964a. The stratigraphy and sedimentary features of the Old Red Sandstone, Portishead coastal section, N.E. Somerset. *Proc. Geol. Ass.*, **75**, 199–221, 7 figs.

1964b. The Triassic Dolomitic Conglomerate and structure of the Old Red Sandstone, Portishead coastal section, N.E. Somerset. *Proc. Bristol Nat. Soc.*, **30**, 445–450, 1 fig., 1 pl.

RAMSBOTTOM, W. H. C. 1970. Carboniferous faunas and palaeogeography of the southwest England region. *Proc. Ussher Soc.*, **2**, 144–157, 6 figs.

RAW, F. 1952. Structure and origin of the Malvern Hills. *Proc. Geol. Ass.*, **63**, 227–239, 6 figs., 1 pl.

READING, H. G. & POLLE, A. B. 1961. A Llandovery shoreline from the southern Malverns. *Geol. Mag.*, **98**, 295–300, 1 fig., 2 pls.

REYNOLDS, S. H. 1907a. A bone cave at Walton, near Clevedon. *Proc. Bristol Nat. Soc.* **1**, 183–187, 1 pl.

1907b. A Silurian inlier in the Eastern Mendips. *Q. Jl geol. Soc. Lond.*, **63**, 217–240, 8 figs., 1 pl.

1908. The igneous rocks of the Bristol district. *Proc. Geol. Ass.*, **20**, 59–65, 1 fig.

1912a. *A Geological Excursion Handbook for the Bristol District.* (1st Edition). 224 pp., 50 figs. Bristol: Arrowsmith.

1912b. Further work on the Silurian rocks of the Eastern Mendips. *Proc. Bristol Nat. Soc.*, **3**, 76–82, 2 figs.

1916a. Further work on the igneous rocks associated with the Carboniferous Limestone of the Bristol district. *Q. Jl geol. Soc. Lond.*, **72**, 23–42, 7 figs.

1916b. Carboniferous Limestone Series of the Area between Clifton and Clevedon. *Proc. Bristol Nat. Soc.*, **4**, 186–197.

1921. *A Geological Excursion Handbook for the Bristol District.* (2nd Edition). 224 pp., 48 figs., Bristol: Arrowsmith.

1929. The Geology of the Bristol District. *Proc. Geol. Ass.*, **40**, 77–103, 10 figs.

1946. The Aust Section. *Proc. Cotteswold Nat. Fld. Club* **29**, 29–39, 2 figs., 4 pls.

REYNOLDS, S. H. & GREENLY, E. 1923. The Old Red Sandstone and Carboniferous Limestone of the Portishead-Clevedon area. *Proc. Bristol Nat. Soc.*, **6**, 92–97.

1924. The geological structure of the Clevedon-Portishead area. *Q. Jl geol. Soc. Lond.*, **80**, 447–467, 10 figs.

REYNOLDS, S. H. & VAUGHAN, A. 1911. Faunal and lithological sequences in the Carboniferous Limestone Series (Avonian) of Burrington Combe (Somerset). *Q. Jl geol. Soc. Lond.*, **67**, 342–392, 12 figs., 4 pls.

RICHARDSON, L. 1901. Mesozoic geography of the Mendip Archipelago. *Proc. Cotteswold Nat. Fld. Club* **14**, 59–69, 1 fig.

1904. *A handbook to the geology of Cheltenham and neighbourhood.* xii + 303 pp., 19 figs., 20 pls., 3 tbls., map.

1907. The Inferior Oolite and Contiguous Deposits of the Bath Doulting District. *Q. Jl geol. Soc. Lond.*, **63**, 383–436, 7 figs., 2 pls.

1909. On the Sections of Inferior Oolite and the Midford-Camerton Section of the Limpley Stoke Railway, Somerset. *Proc. Geol. Ass.*, **21**, 97–100, 3 pls.

1910. The Inferior Oolite and contiguous deposits of the south Cotteswolds. *Proc. Cotteswold Nat. Fld. Club* **17**, 63–136, 5 figs., 7 pls.

1911. The Rhaetic and Contiguous Deposits of West, Mid and part of East Somerset. *Q. Jl geol. Soc. Lond.,* **67**, 1–74, 4 figs., 4 pls.

ROBINSON, P. L. 1957. The Mesozoic fissures of the Bristol Channel area and their vertebrate faunas. *J. Linn. Soc. (Zoo.)* **43**, 260–282, 6 figs., 4 pls.

RUTTER, J. 1829. Delineations of the North Western Division of the County of Somerset, and its antidiluvian Bone Caves, with a geological sketch of the district. 349 pp., 23 pls., map. London: Longman.

SANDERS, W. 1841. Account of a raised sea-beach at Woodspring Hill, near Bristol. *Rep. Brit. Ass. Advmt. Sci.* **1840** *Trans. Sect.,* 102–103.

SAVAGE, R. J. G. & LARGE, N. F. 1966. On *Bigeria accuminata* and the absence of labryinthodonts from the Rhaetic. *Palaeontology,* **9**, 135–41, 1 pl.

SAVAGE, R. J. G. & WALDMAN, M. 1966. *Oligokyphus* from Holwell quarry, Somerset. *Proc. Bristol Nat. Soc.,* **31**, 185–192, 1 fig., 1 pl.

SEAVILL, E. W. 1941. Geology of the Bristol District. *Proc. Bristol Nat. Soc.,* 4th Ser. **9**, 264–74.

SEILACHER, A. 1967. Bathymetry of trace fossils. *Mar. Geol.,* **5**, 413–428, 4 figs., 2 pls.

SHORT, A. R. 1904. On some Rhaetic sections in the Bristol district, with considerations on the mode of deposition of the Rhaetic series. *Q. Jl geol. Soc. Lond.,* **60**, 170–193.

SIMPSON, S. 1951. A new eurypterid from the Upper Old Red Sandstone of Portishead. *Ann. Mag. nat. Hist.,* **4**, 849–861, 4 figs., 2 pls.

SMITH, D. I. & DREW, D. P. 1975. (eds.) Limestones and Caves of the Mendip Hills. 424 pp., 97 figs., 36 pls., 16 tbls. Newton Abbot: David & Charles.

SMITH, S. & REYNOLDS, S. H. 1929. The Carboniferous Section at Cattybrook, near Bristol. *Q. Jl geol. Soc. Lond.,* **85**, 1–8, 2 pls.

SORBY, H. C. 1851. On the microscopical structure of the calcareous grit of the Yorkshire coast. *Q. Jl geol. Soc. Lond.,* **7**, 1–6.

STEAD, J. T. G. & WILLIAMS, B. P. J. 1973. The Pennant Sandstone of Portishead. *Proc. Bristol Nat. Soc.,* **32**, 307–314, 2 figs., 2 pls.

STEVENSON, C. R. 1970. Discussion of paper. 'Triassic palaeogeography of the British Isles'. *Q. Jl geol. Soc. Lond.,* **126**, 81.

STINTON, F. C. & TORRENS, H. S. 1968. Fish otoliths from the Bathonian of Southern England. *Palaeontology,* **11**, 246–258, 14 figs.

TORRENS, H. S. 1968. Some Fuller's Earth Sections in the South Cotswolds *Proc. Bristol Nat. Soc.,* **31**, 429–438, 1 fig.

(ed.) 1969. International Field Symposium on the British Jurassic. Part B, Excursion No. 2, Guide to North Somerset. pp. 2–26, 10 figs.

TRATMAN, E. K. 1963. The hydrology of the Burrington area, Somerset. *Proc. Univ. Bristol Spelaeol. Soc.,* **10**, 22–57, 4 figs., 5 pls.

1975. The Cave archaeology and Palaeontology of Mendip. *In* Smith, D. I. & Drew, D. P. pp. 352–419, 4 figs., 4 pls. q.v.

TRIMMER, J. 1853. On the Southern Termination of the Erratic Tertiaries, and on the Remains of a Bed of Gravel on the Summit of Clevedon Down, Somersetshire. *Q. Jl geol. Soc. Lond.,* **9**, 282–286, 1 fig.

TUTCHER, J. W. & TRUEMAN, A. E. 1925. The Liassic rocks of the Radstock District (Somerset). *Q. Jl geol. Soc. Lond.,* **81**, 595–666, 17 figs., 4 pls., tbls.

VAN DE KAMP, P. C. 1969. The Silurian volcanic rocks of the Mendip Hills, Somerset; and the Tortworth area, Gloucestershire, England. *Geol. Mag.,* **106**, 542–553, 1 fig., 2 pls.

VAUGHAN, A. 1905. The palaeontological sequence in the Carboniferous Limestone of the Bristol area. *Q. Jl geol. Soc. Lond.,* **61**, 181–307, 5 figs.,

8 pls.

WALDMAN, M. 1965. A Coelacanth, *Macropoma*, from the Chalk of Wiltshire. *Proc. Bristol Nat. Soc.*, **31**, 111–112.

WALLIS, F. S. 1927a. The Old Red Sandstone of the Bristol district. *Q. Jl geol. Soc. Lond.*, **83**, 760–789, 1 pl.

1927b. Notes on sections of Old Red Sandstone in the Bristol district. *Proc. Bristol Nat. Soc.*, **6**, 400–405.

WEBBER, F. J. 1967. Post Depositional Structures in the Lias, South Wales. *J. sedim. Petrol.* **37**, 166–174, 9 figs.

WEBBY, B. D. 1965a. The Stratigraphy and Structure of the Devonian Rocks in the Brendon Hills, West Somerset. *Proc. Geol. Ass.*, **76**, 39–60, 5 figs.

1965b. The Middle Devonian Marine Transgression in North Devon and West Somerset. *Geol. Mag.*, **102**, 478–488, 1 fig.

1965c. The Stratigraphy and Structure of the Devonian Rocks in the Quantock Hills, West Somerset. *Proc. Geol. Ass.*, **76**, 321–344, 4 figs.

WEBBY, B. D. & THOMAS, J. M. 1965. Whitsun Field Meeting: Devonian of West Somerset and Carboniferous of North-East Devon. *Proc. Geol. Ass.*, **76**, 179–194, 5 figs.

WEDLAKE, A. L. & WEDLAKE, D. J. 1963. Some palaeoliths from the Doniford gravels on the coast of west Somerset. *Proc. Som. arch. nat. Hist. Soc.*, **107**, 93–100, 1 fig., 3 pls.

WELCH, F. B. A. 1929. The geological structure of the central Mendips. *Q. Jl geol. Soc. Lond.*, **85**, 45–76, 10 figs., 1 pl.

1933. The geological structure of the eastern Mendips. *Q. Jl geol. Soc. Lond.*, **89**, 14–52, 7 figs., 4 pls.

1956. Note on Gravels at Kenn, Somerset. *Proc. Univ. Bristol Spelaeol. Soc.*, **7**, 137.

WELCH, F. B. A. & TROTTER, F. M. 1961. The Geology of the Country around Monmouth and Chepstow. *Mem. geol. Surv. Gt. Br.*, 155 pp., 11 figs., 2 pls.

WETHERED, E. 1891. The Inferior Oolite of the Cotteswold Hills, with special reference to its microscopical structure. *Q. Jl geol. Soc. Lond.*, **47**, 550–70, 1 pl.

WHITTAKER, A. 1972a. The Watchet Fault – a post-Liassic transcurrent reverse fault. *Bull. geol. Surv. Gt. Br.*, No. 41, 75–80, 1 fig., 2 pls.

1972b. Account of an excursion to the west Somerset coast. *Rep. Trans. Devon. Ass. Advmt. Sci.*, **104**, 200–203.

1973. The central Somerset Basin. *Proc. Ussher Soc.*, **2**, 585–592, 2 figs.

WHITTARD, W. F. 1949. Geology of the Aust-Beachley District, Gloucestershire. *Geol. Mag.*, **86**, 365–76, 1 fig., 3 pls.

WICKES, W. H. 1904. The Rhaetic bone beds. *Proc. Bristol Nat. Soc.*, **5**, 213–225, 1 fig.

1908. Pebble swallowing animals. *Proc. Bristol Nat. Soc.*, **2**, 25–31.

WILSON, R. C. L. 1966. Silica diagenesis in Upper Jurassic limestones of southern England. *J. sedim. Petrol.*, **36**, 1036–49, 5 figs.

WITCHELL, E. 1882. *The geology of Stroud and the area drained by the Frome.* 108 pp., 3 pls. Stroud: Geo. H. James.

WOODWARD, H. B. 1876. Geology of East Somerset and the Bristol Coal-fields. *Mem. geol. Surv. England and Wales.* 271 pp., 23 figs., 9 pls.

1893. The Jurassic rocks of Britain. vol. 3. The Lias of England and Wales (Yorkshire excepted). *Mem. geol. Surv. U.K.* xii + 399 pp., 89 figs., 1 map, tbls.

1894. The Jurassic rocks of Britain. vol. 4. The Lower Oolitic rocks of England (Yorkshire excepted). *Mem. geol. Surv. U.K.* 628 pp., 137 figs., 2 pls.